FREE-LIVING AMEBAS:

Natural History, Prevention, Diagnosis, Pathology, and Treatment of Disease

Author

A. Julio Martinez, M.D.

Professor of Pathology
University of Pittsburgh
School of Medicine
and
Neuropathologist
Presbyterian-University Hospital
Pittsburgh, Pennsylvania

CRC Press, Inc.
Boca Raton, Florida

Library of Congresss Cataloging in Publication Data

Martinez, A. Julio (Augusto Julio)
    Free-living amebas.

    Includes bibliographies and index.
1. Amoeba.   2. Amebiasis.   I. Title.   [DNLM:
1. Amebiasis.   2. Amoeba.   3. Amoeba--pathogenicity.
QX 55 M385f]
QR201.A55M37   1985   616.8'0416   84-23854
ISBN 0-8493-6631-3

Direct all inquiries to CRC Press, Inc., 2000 Corporate Blvd., N.W., Boca Raton, Florida, 33431.

© 1985 by CRC Press, Inc.

International Standard Book Number 0-8493-6631-3

Library of Congress Card Number 84-23854
Printed in the United States

FOREWORD

Malaria, amebic dysentery, diseases caused by the trypanosomes; these are some of the better-known protozoal diseases that afflict humans and take a huge toll in lives every year. By comparison, the diseases caused by free-living amebas are insignificant. Yet their virulence and the lack of appropriate therapy have given them an impact far exceeding their incidence. Unlike most protozoal infections, inadequate sanitation or insect vectors are not factors in their occurrence. The organisms exist in nature, their numbers enhanced to some extent by thermal pollution and disturbance of the ecological balance of lakes and rivers. These amebas have an uncanny ability to survive either as free-living organisms or endozoic disease agents, a facultative capacity not usually found among pathogens.

Fortunately, not all free-living amebas are potential pathogens. Pathogenicity appears to be associated with several species in the genera *Naegleria* and *Acanthamoeba*. Though they have been known since the last century and are widely employed in biochemical, nutritional, and morphogenetic studies, the disease potential of small free-living amebas has only recently been recognized. Long known as part of the soil microfauna, their role in diseases of the central nervous system of humans came as a surprise even to biologists studying these organisms. In the case of primary amebic meningoencephalitis caused by *Naegleria,* the disease occurs in otherwise healthy youngsters exposed to the amebas during swimming or other water activities. Onset of the disease is rapid, diagnosis complicated by lack of familiarity with the etiological agent, and the chemotherapeutic arsenal is limited. *Acanthamoeba* encephalitis presents a somewhat different picture, occurring as it does in debilitated or compromised individuals as an insidious illness with a poorly defined portal of entry. Again, effective chemotherapy is lacking.

This text explores the background of these pathogenic amebas, methods for working with them in the laboratory, and basic procedures for their identification, as well as their role in the two different types of amebic diseases of the central nervous system. At one time these organisms were of little concern to those outside of the field of protozoology. But things are different now. Their role in disease has made it important that microbiologists, clinicians, pathologists, and parasitologists become familiar with these amebas, if only to be able to recognize them in samples of cerebrospinal fluid and tissue sections. For these reasons, the text will be much appreciated by those anxious to increase their understanding of these free-living amebas or who, because of circumstances, must have a crash-course in these organisms. Beyond that, it is to be hoped that this book will stimulate readers to examine many of the unanswered questions that exist about these amebas and their role in human disease.

Frederick L. Schuster

PREFACE

During the last decade there has been an increased interest in small free-living, amphizoic, or *limax* amebas because it was discovered that *Naegleria fowleri* and various species of *Acanthamoeba* are pathogenic and capable of producing fatal meningoencephalitis in man and animals. It has also been found that *Naegleria australiensis*, *Naegleria jadini,* and *Naegleria lovaniensis* are not pathogenic even though they are morphologically undistinguishable from *Naegleria fowleri.* It was already well known that amebiasis caused by the parasitic *Entamoeba histolytica* is one of the most serious and devastating diseases in the tropical and semitropical regions of the world. However, the discovery that free-living amebas may produce fatal infections was a great surprise and provoked intense interest in this field of parasitology.

Many factors make protozoology and the study of free-living and parasitic amebas a subject of increasing interest for the medical profession. First, the world has become smaller because of significant advances in technology and transportation, thus protozoa were able to travel faster, and to unusual places. Second, the pollution of water, air, soil, and other natural resources has become a tragic testimony of the deterioration of our environment. Third, the changes in social patterns and alterations in human defense mechanisms have created a more favorable environment for parasites and saprophitic organisms to become opportunistic pathogens. Therefore, infections in immunocompromised hosts are rapidly developing, and new diseases are being ascribed to microorganisms considered until recently non-pathogenic.

Medical students and physicians, medical technologists, nurses, microbiologists, and public health personnel should have a basic knowledge in the epidemiology, prevention, diagnosis, and treatment of diseases produced by free-living amebas. One of the aims of this book is to provide such knowledge and stimulus for research and teaching in this field of medical parasitology and to convince some skeptical individuals that free-living amebas are suitable tools for solving problems of general biological interest.

This book places greater emphasis on the practical problems of epidemiology, prevention, diagnosis, and pathological aspects of the two principal types of free-living amebic infection: the acute disease produced by *Naegleria fowleri* and the chronic, subacute and opportunistic form, produced by *Acanthamoeba* spp.

The text proceeds from basic and general protozoological aspects of free-living amebas and their history to the sequence of events in human disease, i.e., epidemiology, ecology, prophylaxis, primary infection, and specific clinical syndromes, with discussions of the still difficult task of therapy. It continues with pathological features and laboratory diagnosis and concludes with the animal model and spontaneous infections in animals. The bibliography provides additional sources of information regarding other free-living, amphizoic and parasitic amebas and the diseases they cause, including some historical and taxonomic aspects.

There is no doubt that amebologists, protozoologists and other inquisitive minds will find some mistakes, impardonable gaps and perhaps many complicated and surprising formulations and inaccurate statements. However, I hope that this book will not receive the comments made by the British writer Samuel Johnson after reading the Cadogan's Dissertation on the Gout, when he said "all that is good he stole, all the nonsense is evidently his own", which is basically true in all books because "only lunatics can be completely original".

This book is not the work of just one person or ⬛⬛⬛ persons, it is the fruit of generous cooperation, generosity, assistance, friendships, a ⬛⬛ criticism. Especially deserving of thanks are the many pathologists and neuropath⬛⬛ists who sent me formalin-fixed, paraffin-, or plastic-embedded tissues for morphological studies. A special sense of gratitude is felt for the late Dr. Eddy Willaert, a good friend, a pioneer in the study of free-living amebas and a remarkably enthusiastic scientist. A decisive role in the development of this book was that played by John Moossy, M.D. who supported, guided, and offered counsel; to him my sincere thanks. My appreciation and gratitude are also given to Dr. E. Clifford Nelson, Professor Emeritus at M.C.V. and now retired, and to Dr. David T. John, now at Oral Roberts University in Tulsa; they stimulated my interest in the subject and gave me valuable information. I especially thank Dr. John G. dos Santos, Neto, Medical College of Virginia in Richmond, for introducing me to the free-living amebas, and for his advice, encouragement, and editorial assistance; he also supplied valuable clinical and laboratory information on PAM cases from Virginia. My thanks to Dr. George P. Vennart, Professor and Chairman, Department of Pathology, and to Dr. William I. Rosenblum, Neuropathologist, for allowing me to use the files of the Medical College of Virginia in Richmond. Special gratitude and sincere thanks are extended to Dr. Frederick L. Schuster, Professor of Biology at Brooklyn College, for his exceptional editorial assistance; he read the entire manuscript and made substantial and sensible clarifications.

Particular thanks also to Dr. E. J. Cooper, Acting Chief Executive Officer, and to Dr. A. S. Cameron, Senior Medical Officer of the Communicable Disease Control Unit of the South Australian Health Commission, and to Dr. M. M. Dorsh, for valuable information about the public compaign for prevention of PAM in Australia; to Dr. Tim J. Brown, Reader in Microbiology, Massey University, Palmerston North, New Zealand, for valuable advice and suggestions and for the map of New Zealand; and to Drs. David T. John and Thomas B. Cole, Jr., for their advice and assistance, as well as for the use of their scanning electron micrographs.

I would like to thank sincerely Drs. Johan De Jonckheere, Govinda S. Visvesvara, Klaus Janitschke and Witold Kasprzak for sharing with me their valuable knowledge of isolation, culture, and protozoological characteristics of aerobic free-living amebas. I am indebted to Mrs. Margaret Boring and Mrs. Marie LeRoy for the excellent quality of the histologic preparations. To Miss Agnes C. Zachoszcz and Miss Karen M. Perkins I extend my deepest gratitude for their patience, good humor, and excellent secretarial skills. To Miss Linda A. Shab, my gratitude for assistance in photography and superb graphic art work.

To Josephine B., Killeen Jo., Bridget Elizabeth, and Mary Ondina for their encouragement, support, and understanding, I offer my sincerest and warmest gracias...muchas gracias.

This is not the end. It is not even the beginning of the end. But it is, perhaps, the end of the beginning...

Sir Winston Churchill

THE AUTHOR

Augusto Julio Martinez, M.D., was born in Cuba. After earning his undergraduate degree at the Institute of Secondary Education of Camaguey, he received his M.D. from the University of Havana in 1959. He completed postgraduate training in Pathology at the University of Tennessee and Neuropathology at Case Western Reserve University. He was associate professor of neuropathology at the University of Tennessee and the Medical College of Virginia. In 1976 he was appointed to the Faculty at the University of Pittsburgh School of Medicine where he is currently a Professor of Pathology (Neuropathology) and the neuropathologist at Presbyterian-University Hospital in Pittsburgh.

Dr. Martinez is a member of the American Association of Neuropathologists, the International Academy of Pathology, the College of American Pathologists, the American Academy of Neurology, the American Society of Clinical Pathologists, the Society of Protozoologists, the American Association of Pathologists, the Royal Society of Tropical Medicine and Hygiene; he also has been awarded honorary membership in the Sociedad Boliviana de Patología, Associación Peruana de Patólogos, Sociedad Venezolana de Anatomiá Patológica and the Associación de Parasitólogos Españoles. He has served on the Editorial Board of the Journal of Neuropathology and Experimental Neurology and the Morfologia Normal y Patológica.

He has over 100 published articles and abstracts in the field of Neurology, Pathology, and Neuropathology.

TABLE OF CONTENTS

Chapter 5

The Disease: Clinical Types — Manifestations and Course.............................63

Chapter 6

Anatomo-Pathological Characteristics: PAM and GAE 77

Chapter 7

Laboratory Diagnosis: PAM and GAE — Cerebrospinal Fluid Examination........ 105

Chapter 1

INTRODUCTION

I. GENERAL BACKGROUND OF PATHOGENIC PROTOZOA AND FREE-LIVING AMEBAS AND THE DISEASES THEY PRODUCE — REASONS TO STUDY THEM

Why write a book about free-living amebas and the diseases they produce? Why study free-living protozoa and how they are linked to human disease? Why study and spend time and effort on a relatively small group of protozoa which have no apparent economic, medical, or veterinary significance? First, some free-living amebas are pathogenic to man and animals. Second, free-living amebas could behave as facultative parasites to domestic animals and human beings. Third, free-living amebas are eukaryotic cells and can be used as research tools for pathologists, cytologists, cell biologists, protozoologists, geneticists, and electron microscopists. Fourth, studying these fascinating unicellular creatures may bring some insights into many unknown biologic processes. Finally, free-living amebas and other protozoa have had an important influence on human history and geopolitics. Some nations remain underdeveloped largely due to tl.e harmful protozoan diseases which are still prevalent among its citizens and domestic and wild animals. As such, the mechanisms for spread and transmission, epidemiology, and control of protozoan diseases remain an important public health problem.

Free-living amebas are the most recently discovered protozoa that can produce lethal effects on human beings or domestic animals.[4,13,15,17,22,27,43] They can invade the central nervous system and other organs, causing death or permanent disability.[2,3,9,30,31,40,46] They were considered saprophitic and nonpathogenic until Culbertson first established the pathogenic potential and characteristics of *Acanthamoeba* spp. in 1958.[13-16] Free-living or amphizoic amebas of the genera *Naegleria* and *Acanthamoeba* may be either facultatively endozoic (or parasitic), facultatively free-living, or opportunistic.[21,26,42] However, Page used the term "amphizoic" to describe those forms that can exist freely in nature or as endozoic symbiosity.[42] Perhaps all free-living amebas have endozoic yearnings, but this has not been demonstrated beyond *Naegleria* and *Acanthamoeba.*[55] The discovery that the presumably harmless free-living amebas can produce a rapidly fatal meningoencephalitis "has revolutionized the traditional concept of parasitism."[50] It now seems incredible that one of the most inexorably fatal infections of man should escape detection until 1965 when the first human case of primary amebic meningoencephalitis (PAM) was reported by Fowler and Carter from Australia.[22] Almost simultaneously, other human cases of meningoencephalitis due to *N. fowleri* were reported from the U.S.[5] and Czechoslovakia,[10] confirming the findings of Culbertson.[13,14] A significant amount of knowledge has been accumulated since 1958 after the observations of Culbertson and collaborators and since the report of Fowler and Carter in 1965. Much of the now basic information about free-living amebas, their characteristics, and the diseases they can produce has been properly described and reported in multiple clinical, pathological, and protozoological review articles and book chapters.[1,5-9,11-13,15-17,19-21,23-29,32-38,41-45,48,49,51-54]

Unlike *Entamoeba histolytica*, a true parasite which may produce brain, lung, or liver abscesses[4,39] from a primary focus in the colon, pathogenic amebas of the genera *Naegleria* and *Acanthamoeba* can produce fatal CNS involvement, referred to as primary amebic meningoencephalitis (PAM)[3] and granulomatous amebic encephalitis (GAE), respectively.[34] Cases reported from all continents indicate that these amebas are widespread all over the world and suggest that additional cases must have been overlooked. *E. histolytica* is more common in the tropics and subtropics than in the temperate zones, while free-living amebas are ubiquitous. Approximately 10% of the world population harbors *E. histolytica*. The prevalence in the U.S. is less than 3%. The illness produced by this particular ameba has been associated with poor sanitation and poverty. In some developing countries up to 40% of the population may be infected. The infection may occur after ingesting the cysts, usually through food and water fecally contaminated with raw sewage.[39]

Free-living amebic infection, particularly caused by *N. fowleri,* has occurred in young, healthy individuals,[5,9,46] while *Acanthamoeba* spp. has usually occurred in chronically ill, debilitated individuals — affecting lung, skin, and brain.[33-36] Acanthamebic keratitis usually affects healthy individuals.[34] Therefore, the public health implications pertaining to free-living amebas and other pathogenic protozoa are obvious.

REFERENCES

1. Akai, K., Martinez, A. J., and Nakamura, T., Primary amebic meningoencephalitis and granulomatous encephalitis due to free-living amebas. Protozoology, epidemiology and neuropathology (Japanese), *Neurol. Med.-Chir. (Tokyo),* 12, 75, 1980.
2. Anon., Amoebic meningoencephalitis, *Med. J. Aust.,* 1, 1036, 1969.
3. Anon., Primary amoebic meningoencephalitis, *Br. Med. J.,* 1, 581, 1970.
4. Brandt, H. and Tamayo, R. P., Pathology of human amebiasis, *Hum. Pathol.,* 1, 351, 1970.
5. Butt, C. G., The pathology of amebic encephalitis, *Bull. Pathol.,* 9, 83, 1968.
6. Byers, T. J., Growth, reproduction and differentiation in *Acanthamoeba,* in *International Review of Cytology,* Bourne, G. H. and Danielli, J. F., Eds., Academic Press, New York, 1979, 283.
7. Carosi, G., Filice, G., Scaglia, M., Gatti, S., and Torresani, P., Electron microscope study of *Acanthamoeba castellanii*-group spp. A. castellanii, A. rhysodes, A. polyphaga, *Riv. Parassitol.,* 39, 49, 1978.
8. Carter, R. F., Description of *Naegleria* sp. isolated from two cases of primary amebic meningoencephalitis and of the experimental pathological change induced by it, *J. Pathol.,* 100, 217, 1970.
9. Carter, R. F., Primary amoebic meningoencephalits. An appraisal of present knowledge, *Trans. R. Soc. Trop. Med. Hyg.,* 66, 193, 1972.
10. Červa, L. and Novak, K., Amoebic meningoencephalitis: sixteen fatalities, *Science,* 160, 92, 1968.
11. Chang, S. L., Small, free-living amebas: cultivation, quantitation, identification, classification, pathogenesis and resistance, *Curr. Top. Comp. Pathobiol.,* 1, 201, 1971.
12. Chang, S. L., Etiological, pathological, epidemiological and diagnostical considerations of primary amebic meningoencephalitis, *Crit. Rev. Microbiol.,* 3, 135, 1974.
13. Beaver, P. C., Jung, R. C., and Cupp, E. W., Eds., *Clinical Parasitology,* 9th ed., Lea & Febiger, Philadelphia, 1984.
14. Culbertson, C. G., Smith, J. W., and Minner, J. R., *Acanthamoebae:* observations on animal pathogenicity, *Science,* 127, 1506, 1958.
15. Culbertson, C. G., Smith, J. W., Cohen, H. K., and Minner, J. R., Experimental infection of mice and monkeys by *Acanthamoeba,* *Am. J. Pathol.,* 35, 185, 1959.
16. Culbertson, C. G., Pathogenic *Naegleria* & *Hartmannella* (*Acanthamoeba*), *Ann. N.Y. Acad. Sci.,* 174, 1018, 1970.
17. Culbertson, C. G., The pathogenicity of soil amebas, *Annu. Rev. Microbiol.,* 25, 231, 1971.
18. Cursons, R. T. M. and Brown, T. J., Identification and classification of the etiologic agents of primary amebic meningoencephalitis, *N.Z. J. Mar. Freshwater Reserve,* 10, 245, 1976.
19. Duma, R. J., Amoebic infections of the nervous system, in *Handbook of Clinical Neurology,* Vol. 35, Vinken, P. J. and Bruyn, G. W., Eds., North-Holland, Amsterdam, 1979, 25.
20. Duma, R. J., Primary amebic meningoencephalitis, *CRC Crit. Rev. Clin. Lab. Sci.,* 3, 163, 1972.
21. Fernández-Galiano, F. D., Las amebas anfizoicas del hombre, Monograph, Instituto de España, Real Academia Nacional de Medicina, opening ceremony of 1979 course, Madrid, Spain, 1979.
22. Fowler, M. and Carter, R. F., Acute pyogenic meningitis probably due to *Acanthamoeba* spp.: a preliminary report, *Br. Med. J.,* 2, 740, 1965.

23. Fulton, C., Amebo-flagellates as research partners: the Laboratory biology of *Naegleria* and *Tetramitus*, in *Methods in Cell Physiology 4*, Prescot, D. M., Ed., Academic Press, New York, 1970, 341.

24. Fulton, C., Macromolecular syntheses during the quick change act of *Naegleria*, *J. Protozool.*, 30, 192, 1983.

25. Griffin, J. L., Pathogenic free-living amoebae, in *Parasitic Protozoa*, Vol. 3, Kreier, J. P., Ed., Academic Press, New York, 1978, 507.

26. Guevara-Pozo, D., Amebas Limax, Monograph, opening ceremony of 1979 course, University of Granada, Granada, Spain, 1979, 1.

27. Jadin, J. B., De La méningoencéphalite amibienne et du pouvoir pathogène des amibes "Limax", *Ann. Biol. (Paris)*, 12, 305, 1973.

28. Jadin, J. B., Les amibes dans les eaux, *Pathol. Biol. (Paris)*, 22, 81, 1974.

29. Jeon, K. W., *The Biology of Amoeba*, Jeon, Kurang, W., Ed., Academic Press, New York, 1973.

30. Jirovec, O., Parasitisme artificiel des protozoaires libres, *Ann. Parasitol. (Paris)*, 42, 133, 1973.

31. Jirovec, O., Les amibes du type "Limax" comme agent vecteur des meningoencephalities chez l'homme, *J. Med. Lyon*, 50, 1701, 1969.

32. John, D. T., Primary amebic meningoencephalitis and the biology of *Naegleria fowleri*, *Annu. Rev. Microbiol.*, 36, 101, 1982.

33. Lockey, M. W., Primary amoebic meningoencephalitis, *Laryngoscope*, 88, 484, 1978.

34. Martinez, A. J. and Amado-Ledo, D. E., Meningoencefalitis y encefalitis producidas por amebas de vida libre. Protozoología, epidemiología y neuropathología, *Morfol. Normal Patológ.*, 3(Sect. B), 679, 1979.

35. Martinez, A. J. and Janitschke, K., Amobenzephalitis durch *Naegleria und Acanthamoeba*. Vergleich und gegenuberstellung der organismen und der Erkrankungen, *Immun. Infek.*, 7, 57, 1979.

36. Martinez, A. J. and Kasprzak, W., Patogenne pelzaki wolnozyjaceprzeglad, *Wiadomosci Parazytol.*, 26, 495, 1980.

37. Martinez, A. J. and DeJonckheere, J. F., Les infections por les amibes libres, *Bull. Inst. Pasteur*, 79, 171, 1981.

38. Martinez, A. J., Free-living amoebae: pathogenic aspects. A review, *Protozool. Abstr.*, 7, 293, 1983.

39. Martinez-Palomo, A., *The Biology of Entamoeba histolytica*, Research Studies Press, New York, 1982.

40. Neva, F. A., Amoebic meningoencephalitis. A new disease?, *N. Engl. J. Med.*, 19, 450, 1982.

41. Page, F. C., An illustrated key to freshwater and soil amoebae with notes on cultivation and ecology, in *Freshwater Biological Association Scientific Publication #34*, Ferry House, Ambleside, Cumbria, 1976.

42. Page, F. C., *Rosculus ithacus* Hawes, 1963, *Amoebida, Flabelluidea* and the amphizoic tendency in amoebae, *Acta Protozool.*, 13, 143, 1974.

43. Payne, J. I., Amoebic meningoencephalitis, *Science*, 161, 189, 1968.

44. Rondanelli, E. G., Carosi, G., Minoli, L., and Filice, G., Le meningoencefaliti amebiche primarie (M.A.P.) da amebe del gruppo *Hartmannella-Naegleria*. Un capitolo nuovo di pathologia amebica, *Terapia*, 57, 136, 1972.

45. Rondanelli, E. G., Carosi, G., Filice, G., and Scaglia, M., Attualita in tema di amebiasi: quadri ultrastrutturali e nosografia dell'amebiasi da *E. histolytica* e delle meningoencefaliti da amebe "a vita libera" pathogene per l'uomo, *Basi Biol. Med. Mod.*, 3, 305, 1980.

46. Saygi, G., *Naegleria gruberi*. A pathogen?, *Lancet*, 2, 273, 1969.

47. Scaglia, M., Stroselli, M., Grazioli, V., Gatti, S., Bernuzzi, A. M., and DeJonckheere, J. F., Isolation and identification of pathogenic *Naegleria australiensis (Amoebida, Vahlkampfiidae)* from a spa in northern Italy, *Appl. Environ. Microbiol.*, 46, 1282, 1983.

48. Schuster, F. L., Small amebas and ameboflagellates, in *Biochemistry and Physiology of Protozoa*, 2nd ed., 215, Lewandowsky, M. and Hutner, S. H., Eds., Academic Press, New York, 1979.

49. Singh, B. N., *Pathogenic and Nonpathogenic Amoebae*, Halstead Press, New York, 1975.

50. Singh, B. N. and Hanumaiah, V., Temperature tolerance of free-living amoebae and their pathogenicity to mice, *Ind. J. Parasitol.*, 1, 71, 1977.

51. Strauss, R. A., Primary amebic meningoencephalitis, *Chicago Med. Sch. Q.*, 31, 30, 1972.

52. Thong, Y. H., Primary amoebic meningoencephalitis: fifteen years later, *Med. J. Aust.*, 1, 352, 1980.

53. Weisman, R. A., Differentiation in *Acanthamoeba castellanii*, *Annu. Rev. Microbiol.*, 30, 189, 1976.

54. Willaert, E., Primary amoebic meningoencephalitis: a selected bibliography and tabular survey of cases, *Ann. Soc. Belge Med. Trop.*, 54, 416, 1974.

55. Schuster, F. L., personal communication, May 7, 1984.

Chapter 2

HISTORICAL PERSPECTIVE

I. CHRONOLOGICAL PERIODS IN THE EVOLUTION OF KNOWLEDGE OF FREE-LIVING AMEBAS

The evolution of knowledge of medical protozoology and the history of the basic discoveries in free-living and parasitic amebas can be divided into specific periods of time, giving the reader an overall historical perspective of the events leading to our present state of knowledge. We should keep in mind, however, that these periods are somewhat capricious and arbitrary and are chosen only to categorize a continuum into understandable parts. The evolution of this knowledge is patched with provocative ideas, exciting theories, innovative hypotheses, and startling discoveries which are significant examples of human curiosity. These advances in knowledge, positive accomplishments, and achievements stimulated studies and new discoveries, leading to the clarification of the role played by free-living or amphizoic amebas in human and animal infections.

A. Period I: 1674 to 1774 (100 years)

This period covered a century. The science of microbiology was born in this era. Antonie van Leeuwenhoek (1632 to 1723), in the summer of 1674, was the first person to observe free-living protozoa under the microscope in a drop of water; and for this reason he should be considered the founder of protozoology and bacteriology. Rösel von Rosenhof, in 1755, made the first observation of a free-living ameba through the microscope.[88] He called it "small proteus". Later, C. Linnaeus classified it as *Chaos proteus*.

B. Period II: 1775 to 1849 (75 years)

This period covered a span of 75 years in which very fundamental discoveries were made by pioneers in the field of protozoology. In 1838, Ehrenberg published his work on the infusoria and created the genus *Amoeba* in Germany.[50] Two years later in 1841, Dujardin, in France, described for the first time "Limax amebas".[47] In 1849 Gros, in Russia, described a parasitic ameba found in the oral mucosa and named it *Amoeba gingivalis*.[54] Later it was named *Entamoeba gingivalis*.

C. Period III: 1850 to 1899 (50 years)

This period covered a span of 50 years. In 1875 Fedor Aleksandrovich (Losch) (1840 to 1903) in St. Petersburg (Leningrad) discovered *Entamoeba histolytica* in the fecal specimens of a 24-year-old man with chronic dysentery. An autopsy revealed extensive ulcerative lesions containing numerous amebas in the colonic mucosa, chiefly in the sigmoid flexure and the descending colon. The freshly passed ameba-containing stools from the patient were inoculated into four dogs who developed similar symptoms and lesions in the colonic mucosa.

Leidy (of the University of Pennsylvania, in 1879) and E. Penard (in France, in 1890) made more precise descriptions of Limax amebas.[68]

In 1892, Flexner reported an abscess of the jaw, probably due to free-living amebas, in a 60-year-old woman from Virginia.[52]

In 1899, Schardinger described a "free-living ameba" in Vienna. He named the organism *Amoeba gruberi* and observed that it was an amebo-flagellate.[90]

D. Period IV: 1900 to 1949

This period covered a span of 50 years. Great achievements were accomplished during this period. In 1900, Dangeard in France studied and described the mitosis of the *Amoeba hyalina* and stated that "nothing is more difficult, really, than classifying an ameba."[38]

In 1905, Vahlkampf described limax amebas in Germany.[99] In 1909, Nagler described the morphological features of the *A. hartmanni*.[74] Hartmann described a parasitic ameba in Germany.[55] In 1911, Alexeieff described the mitosis of limax amebas and established the genera *Hartmannella* and *Naegleria*.[1-4] The term "limax" amebas was used for all small amebas which move in a slug-like motion. Awerinzew (in St. Petersburg, in 1912) applied the term amebo-flagellate to the free-living amebas that can transform into the flagellate state.[11] This term is especially applicable to *Naegleria* spp.

In 1912, Alexeieff proposed the generic name *Naegleria* for limax amebas having the ability to transform into a temporary flagellate phase and having nuclear division characterized by "promitosis". The species *A. gruberi* became the type species.[3] Chatton and LaLung-Bonnaire (in France, in 1912) also described the genus *Vahlkampfia*, noting the absence of a flagellate stage.[29,30]

Prowazek described *Iodamoeba butschlii* and *Entamoeba hartmanni* in 1912.[83] Hogue, in 1914, described limax amebas.[58] Zulueta (in Spain, in 1917) described more precisely the nuclear division of limax amebas.[104]

The terms "soil" amebas and "soil" protozoa came into wide usage in the early 1900s. The name is actually a misnomer, in that these protozoa require water for activity and growth. Only those amebas capable of assuming a resistant "cyst" stage survive drought and become active when in an aqueous environment. "Soil" protozoa such as *Naegleria* and *Acanthamoeba* are common and abundant in fresh water.[76,94,97]

The turn of the century was a period of great activity in determining causative organisms of diseases and led to the use of the terms "coprozoic" ameba and "coprozoic" protozoa. It was found that an ameba was the cause of a fatal form of "tropical dysentery" which eventually became known as amebic dysentery. After more than 20 years of trying to grow the ameba in cultures, Boeck and Drbohlav succeeded in 1925.[12] An axenic medium was devised later.[44] Prior to 1925, there were numerous reports of growing amebas and other protozoa from feces inoculated into culture media, which eventually proved to be nonparasitic or free-living protozoa. For the most part, they were found to be species already commonly known as "soil protozoa". These species, prone to grow out of diluted feces and presumably coming from cysts thought to be in the food and water that was swallowed, survived the digestive process through the intestines. Because of that, these amebas were known as the "coprozoic protozoa". Among these species were *Naegleria*, *Acanthamoeba,* and *Hartmannella*.[39]

During this period of active investigation in separating the parasitic protozoa from the free-living, some instances of parasitism were discovered. For example, in 1921 it was reported that *V. patuxent*, a limax ameba, was able to live as a parasite in the digestive tract of oysters.[59] In 1926 in the U.S., Schaffer published a monograph describing and classifying fresh-water and marine amebas.[89]

By 1930, interest in the limax amebas was in decline. The role in disease had been reduced to the status of free-living coprozoic amebas and even significance as soil organisms were limited to interest by a few devoted students in agriculture. Bacteriologic studies played a key role in the hospital and public health laboratories investigating infectious disease organisms. Bacterial cultivation on solid agar plates and slants had become a standard procedure. The year 1930 is memorable because on that date, Castellani reported a limax ameba growing on yeast and bacteria previously inoculated into an agar plate.[21-24] Castellani, intrigued by a remarkable clearing of bacteria and yeasts on his plate, made an intensive study of the ameba. Douglas, in 1930, stimulated by this report of a laboratory contamination by an ameba, attempted to classify it, naming this ameba *H. castellanii.*[46] The agar plate method of culture on bacteria was widely adopted for research on the biology of these amebas.[75]

Of particular interest was the discovery of morphological evidence justifying the establishment of a new genus for the ameba found by Castellani.[25,100] In 1931, Volkonsky differentiated it from other harmannellid amebas and gave it the name *A. castellanii.*[100] Workers in the field devoted their efforts to establishing criteria for dividing the limax amebas into two major groups. The genera *Hartmannella* and *Acanthamoeba,* characterized by similar features such as mitotic nuclear division, were commonly grouped as the hartmannellid amebas. The amebo-flagellate, *Naegleria,* and species of *Vahlkampfia* were classified in the family Vahlkampfiidae because they share promitotic nuclear division and other common features of the family Vahlkampfiidae.[60,79,96] Shinn and Hadley, in Pittsburgh, noted spontaneous contamination of a bacterial culture by *H. castellanii* in 1936;[91] and Hewitt, in 1937, also reported free-living amebas contaminating bacterial cultures.[57] Singh (in India, in 1941) studied soil amebas in pure and mixed culture.[95] Rafalko, in 1947, used the term amebo-flagellate on *N. gruberi* because it can be transformed temporarily into a flagellate form.[84]

E. Period V: 1950 to Present (1984)

This period is still evolving. During the 34 years from 1950 to 1984, significant milestones were accomplished; and this period is considered to be the most productive in the history of free-living amebas. Singh, in 1950 and 1952, did fundamental work that formed the basis for the modern taxonomic classification of free-living amebas according to the nuclear division patterns.[95]

In 1950, Nakanishi in Japan isolated free-living amebas in river water from Java.[76] Nakamura, in 1951, described the phagocytic capabilities of free-living amebas to bacterial cultures.[75] In 1951, Asami and Nakamura in Japan classified an ameba which spontaneously contaminated agar-plates of *Shigella* culture.[10] In 1952, Herrera in Panamá erroneously reported a case of a boy who died of amebic encephalitis.[56] Later, this case was reexamined by Johnson who found that, indeed, free-living amebas were the etiological agent and probably *Naegleria fowleri.*[65] In 1952 and 1953, de la Arena, in Cuba, isolated amebas from fresh water, naming one of them *Astramoeba torrei* and placing the other one in the genus *Mayorella* (Schaeffer).[40,41] The same author, in 1955, again isolated a new species of ameba, calling it *A. tatianae.*[42]

In the early 1950s, interest in the limax amebas might have subsided to the routine of basic studies, however, the versatility of the limax amebas again began to invade the virology laboratories.[20,27,28] The perfection of tissue culture techniques as a tool had furthered the development of specialized laboratories for the diagnosis and study of viral infections. In 1956, while doing virology research in the U.S., Jahnes et al. noted spontaneous amebic contamination of monkey kidney tissue culture by *Hartmannella* spp. causing cytopathic effects.[64] In 1959, Chi et al. reported selective phagocytosis of erythrocytes by free-living amebas in cell culture.[31] It was noted that the ameba identified as *Acanthamoeba castellanii* produced plaques singularly similar to those produced by viruses.[49,51,73] However, some studies attributed the "contaminant" to the cell source rather than the throat swab. Later studies by Wang and Feldman in 1961 and 1967,[101,102] Chang et al. in 1966,[27,28] Pereira et al. in 1966,[82] and Armstrong and Pereira in 1967, reported isolations of hartmannellid amebas from nasal, throat, and bronchial fluid.[9] In several instances it was first assumed that a virus was the cause of the cytopathic effect and plaques in the tissue cultures.[10,20,25,32,33,49,82,103] Amebic isolation from throat swabs was definitely more useful than isolation from the lower gastrointestinal tract, and this was simply another way to study "soil" or "dust" amebas.[101,102]

The turning point came in 1958 as a by-product of tissue culture and virus laboratory work. In previous instances, discovery of invasion of laboratory cultures by limax amebas was always missing from the study. Circumstances now introduced the key, a fortuitous instance of animal inoculation which revealed a completely unexpected ability of Limax amebas to destroy the CNS tissue and produce a fatal meningoencephalitis. In 1958, at the Lilly Research Laboratories in Indianapolis, tissue cultures were being used in polio vaccine safety tests.[32] When plaques appeared in the culture of Rhesus monkey kidney cells, Culbertson thought a virus was present. As part of the vaccine safety test routine, mice and monkeys were inoculated intravenously, intracerebrally, and even intranasally. Unexpectedly, all of the animals died. Histopathological study of the tissues and cultures revealed an ameba growing and producing the plaques that originally were thought to be of viral origin. The ameba, a species of *Acanthamoeba*, was then found in the tissue cultures, apparently an airborne contaminant. The discovery that the ameba after simple intranasal instillation in mice could invade the olfactory mucosa, migrate to the brain, and produce a fatal meningoencephalitis was a shocking revelation. Culbertson reported the pathologic findings and suggested that in view of the frequent exposure of people to these common amebas, infection and even disease in human beings should be considered a possibility.[33-35] In 1961, Culbertson received the Ward Burdick Award of the ASCP for demonstrating the pathogenic potentiality of *Acanthamoeba* spp.[33] He predicted the possibility of occurrence of free-living amebic human infection that was demonstrated a few years later.[5-8,13-19,26,48,61-63,66,86,87]

The decade 1958 to 1968 virtually encompasses the complete history from detection to full maturity of knowledge of primary amebic meningoencephalitis (PAM)[62,71,72] and granulomatous amebic encephalitis (GAE).[47,50] The chain of events consists of five major links which served to bring the existence of the disease into view and identify the amebic agent:

1. A fortuitous animal inoculation test which, for the first time, revealed the pathogenicity of a Limax ameba and the production of meningoencephalitis in the laboratory animals[32]
2. Fatal human cases discovered with histopathological features similar to those produced by *Limax* ameba in the laboratory animals[13-19,33]
3. Isolation of amebas from the fatal human cases[15,17,18,35]
4. Identification of a *Naegleria* ameba isolated from patients and grown in culture in the laboratory (a surprise because the initial discovery of the pathogenicity of a Limax ameba involved only a species of *Acanthamoeba*)[77,78]
5. Production of the disease in laboratory animals with similar histopathological changes as in the human cases[32]

The amebic strains from fatal human cases have proven incredibly virulent. Isolation of such strains from natural sources was very difficult. *Acanthamoeba* spp. was apparently not involved in fatal cases of meningoencephalitis in man,[17] and, apparently, the only species involved in the early cases was *Naegleria*.[13,35]

In 1965, the first detailed report of four fatal human cases from Australia was published by Fowler and Carter where the first case occurred in January 1961.[53] These cases were all detected after death by histopathologic studies and resembling the pathological changes produced in animals by Culbertson.[32-34]

The first three fatal cases in the U.S. were reported from Florida in 1965 to 1966 by Butt where the first case occurred in 1962.[13,15] Butt presented a preliminary report in 1964 (Figure 1) at the annual meeting of the ASCP on October 16 to 24, 1964.[13] Ironically, the abstract was not suitable for publication in the *American Journal of Clinical Pathology* in spite of the fact that it conveyed a message of paramount importance

THE AMERICAN JOURNAL OF CLINICAL PATHOLOGY
Copyright © 1964 by The Williams & Wilkins Co.

Vol. 42, No. 5
Printed in U.S.A.

ABSTRACTS OF PAPERS

SCIENTIFIC SESSIONS OF THE ANNUAL MEETING OF THE AMERICAN SOCIETY OF CLINICAL PATHOLOGISTS, OCTOBER 1964

AMEBIC MENINGOENCEPHALITIS

Cecil G. Butt
(Abstract not suitable for redaction
and publication in the JOURNAL)

FIGURE 1. Program of the Scientific Sessions of the Annual Meeting of the ASCP held in Bal Harbour, Fla., October 16 to 24, 1964, announcing the paper presented by Dr. C. G. Butt entitled "Amebic Meningoencephalitis. (From *Am. J. Clin. Pathol.*, 42, 516, 1964. With permission.)

(Figure 1). In two of the cases amebas were noted in the cerebrospinal fluid (CSF) but could not be cultured. From one case an ameba was iosolated by animal inoculation and later identified as a species of *Naegleria*. Butt named the disease Primary Amebic Meningoencephalitis (PAM).[14] Again in Florida, in July 1966, another case with evidence of PAM was seen by Butt.[15] After death, CSF and brain tissue were placed in culture media. Isolation was achieved in the laboratories of both Culbertson who classified it as *Naegleria* spp. (HB-1) and Butt who classified it as *Naegleria* spp. They identified the organism as a species of *Naegleria*.[35]

The reports from the U.S. showed evidence that PAM was not a new disease but had occurred and escaped recognition until that time. In 1968 eight cases of rapidly fatal meningitis occurred in the Richmond area. Seven of these cases occurred in a cluster in the summers of 1951 and 1952. Autopsies were performed on seven of the eight cases. Callicott examined the slides and was able to establish the diagnosis of PAM.[16] Another case on August 15, 1968, prompted the review of all autopsy records from the Medical College of Virginia dating back to 1920 by dos Santos. Not only was it discovered that the seventh case from the period of 1951 and 1952 was PAM, but four more cases were uncovered. This retrospective study revealed that the first case occurred in Virginia on July 15, 1937.[45]

The first isolation with identification of the species of the ameba from CSF before death of a patient was accomplished in Virginia by Nelson and Jones.[77,78] On July 31, 1967, a patient with symptoms suggesting PAM was admitted to the Medical College of Virginia Hospitals. Amebas were noted in the CSF. Culture and isolation was obtained from the CSF before death and after death from brain tissue and CSF. Study of the cultures revealed the ameba to be a species of *Naegleria*. This isolate was called strain MCV-CJ1. The second isolation was done on August 13,1968, and called MCV-LEE1. The third isolation was done on July 11, 1969, and called MCV-TY-1. The fourth successful isolation was performed on July 12, 1969, and called MCV-WM-1. An *A. astronyxis* was also obtained in CSF cultures from another case of meningitis which survived in Richmond, Va.[17] The work done by Drs. E. C. Nelson, M. M. Jones, J. Callicott, and J. G. dos Santos was of fundamental importance in those early days (Figure 2).

A spectacular report, and first from Europe, came in 1968 when Červa reported 16 fatal cases in the years 1962 to 1965 in Czechoslovakia.[26] All were detected after death by the histopathologic, clinical, and epidemiologic features. It is interesting to point out that cases of amebiasis were reported prior to the recognition and establishment of

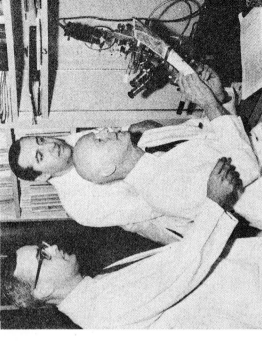

Dr. J. H. Callicott (left), Dr. E. C. Nelson, Dr. J. G. dos Santos
Their Studies of Amoebic Disease Continue at Medical College

Staff Photo

-4 Richmond Times-Dispatch, Sunday, Jan. 21, 1968

15-Year-Old Medical Puzzle Ends

Amoeba Blamed For Epidemics

By Beverly Orndorff
Times-Dispatch
Science Writer

During the summers of 1951 and 1952, a total of nine persons died here as a result of something that appeared to be encephalitis (inflammation of the brain) or a form of meningitis (inflammation of of the brain's coverings).

The victims were from the Richmond area, and from a broad area around Richmond.

Their deaths caused much concern among area physicians, medical scientists at the Medical College of Virginia, and the Richmond Health Department; even the Public Health Service's Communicable Disease Center became involved.

Eastern equine encephalitis, the virus from which is transmitted to humans by mosquito bites, was strongly considered as the cause of the fatal illnesses. Public Health Service officials tended to believe, however, that bacteria, rather than a virus, were involved.

But neither bacteria nor viruses were ever found in the spinal fluids of the victims.

Hence, in the minds of many, the cause of the epidemics was never really discovered. One MCV pathologist, in fact, kept his files on the cases through the years, and occasionally recalled the unsolved mystery with some uneasiness.

Continued From First Page

plates had certain abnormalities thought at first to be caused by viruses. Further studies, however, showed that certain species of amoeba—commonly found in water and soil and capable of being dust borne—were involved.

As a result of those studies, Dr. C. G. Culbertson, director of Lilly's biological research division, suggested the organisms could possibly infect animals, and perhaps even man.

In 1965, two Australian physicians reported the first human cases (four of them) that seemed to fit Dr. Culbertson's descriptions of an amoebic meningoencephalitis. Then came several reported cases in Texas and Florida, and about the same time, the 1966 MCV case was discovered.

This case was a mystery itself, according to the MCV scientists. But by careful study of brain tissue, they were able to identify that case as one of amoebic meningoencephalitis. The case was very similar to those cases reported from Australia, Florida and Texas.

are by no means conquered yet.

In general, the disease has a rapid, usually fatal outcome within about five or six days after onset, and the organism — a species of Naegleria amoebae — seems to be highly resistant to antibiotics and sulfa drugs.

At the same time, there is some suspicion that other species of amoebae — belonging to the group Hartmannella — may also cause central nervous system disease that is milder and not necessarily fatal.

AMONG OTHER UNANSWERED questions are how the organisms infect humans, and what the sources of infections may be.

Both species are commonly found in soil and water, and can be carried by dust. There has been a strong suspicion that the organisms may be contracted from swimming in certain lakes or ponds. But said the MCV scientists, at this point, "We really can't wholly implicate swimming as a source."

Recent evidence, they added, indicates that these amoebae may be found in the nasal area of a certain number of normal individuals.

THEN, IN THE SUMMER of 1966, a patient with signs and symptoms similar to those experienced by the 1951 and 1952 patients, **was** brought into MCV.

That touched off a chain of events leading to the final solution of a 15-year-old medical mystery.

It was neither a virus nor bacteria that **caused the outbreak, but rather an amoeba —a commonly prevalent one-celled organism that had** never been considered to cause brain or central nervous system disease.

A report on all this appears in the January issue of the American Journal of Clinical Pathology. The article was written by Dr. Joseph Callicott, who is in his final **year** of specialty training in pathology at MCV.

He was assisted in his studies of the amoebic meningoencephalitis (it affects both the brain and its coverings) by two medical scientists who were involved in the 1951-52 outbreaks — Dr. E. Clifford Nelson, a microbiologist at MCV, and Dr. J. G. dos Santos, a clinical pathologist there.

Amoebae, they explained in **a recent interview, can cause some diseases in man, but it had been thought they did not directly attack the brain and spinal cord (the central nervous system).**

THE FIRST SUSPICIONS, however, that a certain kind of amoeba might cause central nervous system disorders in animals came during the late 1950s in the course of polio vaccine safety tests at the Eli Lilly pharmaceutical firm.

It was noticed that a particular set of tissue culture

RECOGNITION OF THE 1966 MCV case prompted Dr. Santos to recall the mysterious Richmond epidemic of 1951-52. Re-investigation and review of autopsy material obtained during the 1951-52 epidemic permitted the MCV scientists to recognize the amoebic nature of the brain infections, thereby clarifying the long-standing mystery.

The medical scientists continued their study of amoebic meningoencephalitis, and have subsequently isolated—in Dr. Nelson's laboratory—the offending amoebae. This, they said, will permit further clarification of the nature of such diseases.

Even though, however, those amoebae have been tracked down as the cause of the earlier summer time epidemics, the disease or diseases these organisms cause

And although the cases reported so far have come from Virginia, Texas and Florida, this does not mean that the amoeba-caused central nervous system diseases are necessarily confined to the Southern portion of the nation.

It is possible that the disease is more prevalent, but just hasn't been recognized, according to the MCV medical scientists.

At present, their research is generally aimed at a better understanding of the disease, of the organisms, and, they hope, of developing means of combatting it.

Meanwhile, the first barrier to conquering any disease—isolating and identifying the villains—has been overcome. And that, agreed the MCV scientists, is a hopeful beginning.

FIGURE 2. The pioneers of the study of free-living amebas in the laboratory of Dr. E. C. Nelson at the Medical College of Virginia in Richmond. (From *Richmond Times Dispatch*, January 21, 1968. With permission.)

free-living amebic pathogenicity. Derrick, in 1948, reported a case of free-living amebas originally classified as *Iodamoeba bütschlii* in a Japanese prisoner of war.[43] Later it was identified as *N. fowleri.* Kernohan et al., in 1960, reported a case from Arizona characterized by granulomatous inflammation of CNS tissue, originally thought to be *I. bütschlii,* but probably due to *Acanthamoeba* spp.[67] In 1965 to 1966, Patras and Andujar from Texas reported another case originally believed to be due to *Acanthamoeba* spp. but later recognized by immunoperoxidase technique as *N. fowleri.*[80,81]

In those early years the confusion in terminology regarding the type of ameba was astonishing but understandable.

Mandal et al., in New Zealand,[70] reported the first case from that country erroneously as due to a "slime mold". Later, as a result of the efforts by Brown and his group from Massey University, it was definitely proven to be due to *N. fowleri.*[36,37]

As long as the species involved was in doubt, epidemiology, preventive measures, and even specific therapy remained conjectural. Symmers, in Great Britain, retrospectively reported two cases, one from Essex, east of London, which occurred in 1909, and one case from Belfast which occurred in 1937.[98]

In January 1966, in Adelaide, Australia, Carter again observed a case with symptoms suggestive of (PAM).[18] Amebas were seen in the CSF but were not cultured. After death, cultures inoculated with brain tissue and fluid were successful. Study of the cultures to obtain identification produced a startling finding. The ameba was noted to transform into a temporary flagellate stage. This established the identity as an ameboflagellate, a species of *Naegleria,* a departure from the previous expectations. It has been assumed that the ameba would be a hartmannellid of either the genus *Acanthamoeba* or *Hartmannella.*[15,35] The essential role of culture isolation and organism study for identification was significant.

In 1966, Page in the U.S. established taxonomic criteria for the small free-living amebas with a redefinition of the genus *Hartmannella,*[79] and Pussard in France and Jadin in Belgium made significant contributions regarding taxonomic and epidemiologic features of free-living amebas.[61]

The year 1968, with its reports of successful isolation of the amebic agent of PAM and startling disclosure that it was a species of *Naegleria,* had a surprising and stimulating effect on research programs throughout the world. On the one hand, efforts to obtain isolates from cases for diagnostic and epidemiological purposes were stimulated and serious research efforts were begun to devise an effective therapy. Attention was focused on environmental sources and circumstances of incurring infection. Studies on the possibility of producing disease by species of *Hartmannella* and *Acanthamoeba* were put in an uncertain status. To some extent this was offset by a more vigorous pursuit of cases with evidence of possible involvement of these genera. Finally, a very active inquiry into the taxonomic status of the species of *Naegleria* isolated from human cases ensured. During 1970 cases continued to be identified and the majority were due to *N. fowleri* and very few due to *Acanthamoeba* spp. Kenney in 1971,[66] Robert and Rorke in 1973,[85] and Jager and Stamm in 1972[63] reported the first cases clearly identifying the ameba responsible as *Acanthamoeba* spp., and this definitely implicated the ameba in human cases. Thus, clinicopathological differences between the two types began to emerge. Ocular and skin infections due to *Acanthamoeba* spp. were also found.[69]

On January 20, 1972, Carter presented his memorable paper at the Royal Society of Tropical Medicine and Hygiene in London,[19] and he received the following comment from Professor P. C. C. Garnham: "Breaking a barrier is always a thrill; whether it is the 'sound barrier' or the 'barrier of host parasite specificity', for instance, making human malaria parasites grow in owl monkeys. Dr. Carter's work on the amebo-flagellates provides an example of overcoming an even more formidable obstacle. He has

THE PITTSBURGH PRESS. August 10, 1977.

Texas Lake Swimmer Dies Of Rare Disease

EDINBURG, Tex. (UPI) — A 17-year-old girl who lay in a coma 10 days suffering from a rare disease contracted while swimming with her family died an hour after she was disconnected from a respirator, hospital officials said today.

The girl, Dahlia Reyna, the daughter of Mexican-American migrant farm-workers, died at 11:24 p.m. (EDT) yesterday, according to R. W. Manley, administrator at Edinburg General Hospital.

Hospital officials said Miss Reyna's family, physicians and a priest had conferred about whether the life support system should be disconnected.

Miss Reyna apparently contracted the disease of the nervous system — called meningoencephalitis — while swimming with her family 10 days ago at Delta Lake.

She was admitted in critical condition to the hospital Sunday.

She was kept alive with the aid of a respirator which kept her heart beating, but doctors said she had shown no brain responses since entering the hospital.

The Center for Disease Control in Atlanta said Miss Reyna was only the second known victim to contract meningoencephalitis this year. Only 60 cases of the diseases have been reported in the last 15 years, a doctor said.

The hospital officials said the decision to remove the respirator was made yesterday.

Dr. Charles Marshall, a regional officer for the Texas Department of Health, said the disease is carried by an amoeba which lives in water. He said the tests of Delta Lake water found samples of the amoeba, but he said it had not been determined whether the amoeba was present in dangerous quantities.

Nevertheless, Marshall said swimming has been prohibited at the lake.

No other members of Miss Reyna's family have shown symptoms of the disease.

The family yesterday consulted officials at John Sealy Hospital in Galveston, Tex., in hopes the hospital would accept the girl's case. However, a hospital spokesman said Sealy Hospital would not accept patients who were on a respirator.

FIGURE 3. News about a new case of PAM occurring in Texas. (From *The Pittsburgh Press*, August 10, 1977. With permission of United Press International and *The Pittsburgh Press*.)

shown in one of the most interesting papers presented to the Society how the gulf between free-living and parasitic protozoa can be traversed to give rise to a fulminating infection in the human host.''

Sporadic cases of PAM occurring in several parts of the U.S. were reported in the news media (Figures 3 and 4). Front page headlines of the *Richmond Times Dispatch* on September 3, 1978 looks at the unresolved mystery of cases which occurred 27 years previously and were retrospectively diagnosed by Dr. J. G. dos Santos (Figure 5). In the summer of 1980 another case of an 11-year-old boy from central Florida was reported both in Orlando's Florida area newspaper and elsewhere (Figure 6).

During the decades of 1960 and 1970, several research groups made up of distinguished scientists made significant contributions on free-living amebas and their pathogenic potential; among them are the late Eddy Willaert, J. Jadin, H. Van de Voorde, Johan de Jonckheere, and P. J. van Dijck in Belgium; Klaus Janitschke, S. Dempe, and Helga Jantzen in West Germany; Witold Kasprzak and T. Mazur in Poland; W. Stamm and D. C. Warhurst in England; E. C. Rondanelli, G. Carosi, G. Filice, and M. Scaglia in Italy; M. Proca Ciobanu and G. Lupascu in Romania; B. N. Singh, S. R. Das, S. C. Maitra, and B. W. Krishna Prasad in India; M. Pussard, P. Pernin, C. Derr-Harf, and B. Molet in France; L. Červa, V. Skocil, and C. Servus in Czechoslovakia; S. L. Chang, R. J. Duma, E. C. Nelson, J. G. dos Santos, C. G. Culbertson, J. L. Griffin, F. L. Schuster, T. K. Sawyer, A. R. Stevens, G. S. Visvesvara, T. Byers,

SUNDAY, JULY 30, 1978 .The New York Times.   35

Amoeba That Kills Swimmers Puzzles Medical World

Special to The New York Times

TAMPA, Fla., July 29 — It could be a science-fiction plot: Usually harmless organisms living in fresh water produce a strain that attacks and kills swimmers by feasting on their brain cells.

But it is not fiction. Scientists have discovered just such a creature, an amoeba that almost always kills its victims.

Since the microorganism was identified in 1963, eight deaths in Florida have been attributed to it, as have 17 in Virginia. There have also been victims in South Carolina, Georgia, Texas, Louisiana, New York and California, according to Dr. George Healey, a parasitologist at the Federal Center for Disease Control in Atlanta. Deaths from the amoeba have also been reported in Czechoslovakia, Belgium, New Zealand and Australia.

Several deaths have been attributed to an airborne strain of a similar amoeba, but less is known about this strain.

Hazard May Seem Small

Since the amoeba, called Naegleria fowleri, has been found in many parts of the United States and other countries in which millions of people swim without apparent harm, the hazard it represents would seem to be small.

"Why some people swim in the water and are perfectly all right, and others swim in the same place and get sick, we don't know," said Dr. John Gullett, a specialist in infectious diseases at the University of California in San Francisco.

Scientists studying the organism are puzzled by its potency, its erratic occurrences and its resistance to drugs. The amoeba is usually associated with the sediment on the bottom of warm man-made lakes and sometimes with polluted areas, but it has also been found in cold springs and clean waters.

Its latest confirmed casualty is 14-year-old Philip Eddy, who died on July 9 after swimming in a lake near here. The boy showed symptoms similar to those of flu:

severe headache, nausea and lethargy. He was hospitalized, fell into a coma and did not respond to drugs that had shown some promise in laboratory tests.

Most Harmless to Humans

Most amoebas, which live in great profusion in nature, are harmless to humans. But Naegleria fowleri can enter a swimmer's nose and move up the olfactory nerve to the brain, where it feeds on the tissue.

There have only been 123 cases reported worldwide, and only three survivors, most recently Mary Park, 9, of California who became ill on June 1. Doctors ordered laboratory tests of samples of her spinal fluid, in which a pathologist found signs of the amoeba.

"They thought she had meningitis when Lynne Boyle, the pathologist, recognized the cells were moving," said her doctor, James Seidel, a pediatrician at Harvard General Hospital, which is affiliated with the University of California School of Medicine at Torrance. Doctors immediately prescribed strong antifungicide drugs, one of which is still experimental.

"She was in a coma when she arrived, spent about three days in a deep coma, then she started steadily improving," Dr. Seidel said. She remained in the hospital for a month and has since recovered.

Cases Are to Be Recorded

The Center for Disease Control has begun an international registry to record Naegleria cases, but researchers stress that the condition is easy to miss because of its common symptoms and its similarity to meningitis, an inflamation of the membrane of the spine and brain.

"The problem is the majority of the physicians, neurologists and others are not aware enough of the disease," said Dr. Julio Martinez, associate professor of pathology at the University of Pittsburgh School of Medicine.

Dr. F. M. Wellings, director of the State Epidemiology Research Center here, said, "We believe it is ubiquitous. It is all over the place." She suggested that swimmers wear nose clips or blow their noses after swimming.

FIGURE 4.   News about another case of PAM. (From the *New York Times,* July 30, 1978. With permission.)

and D. T. John in the U.S.; Y. H. Thong, R. F. Carter, A. Jamieson, A. Ferrante, and B. Rowan-Kelly in Australia; Tim J. Brown, R. T. M. Cursons, E. A. Keys, and M. Miles in New Zealand; D. Guevara-Pozo, D. Fernández-Galiano, M. C. Mascaro-Lazcano, and M. J. Madrigal-Sesma in Spain; W. A. M. Linnemans in Holland, and R. C. Brown and T. Brown in Scotland.

In 1980, Martinez, in Pittsburgh, Pa., called the disease produced by *Acanthamoeba* spp. Granulomatous Amebic Encephalitis (GAE)[71] to differentiate it from the PAM due to *N. fowleri*,[72] and again suggested the opportunistic capabilities of *Acanthamoeba* spp. At the beginning of 1984 more than 160 cases of PAM and GAE have been reported from all over the world.

The Weather

Today: Partly cloudy. High in mid-80s. Low tonight in low 60s.
Tomorrow: No change.
Local Data: Page A-8.

Richmond Times-Dispatch

Richmond, Virginia 23219   **Sunday, September 3, 1978**

Virginia's
State
Newspaper

50 Cents

128th Year, No. 246

Deaths in '51, '52 Took 15 Years to Solve

Editor's note: The outbreak of an unknown illness among members of the American Legion who attended a convention in Philadelphia in 1976 became a widely publicized medical mystery. The cause, a never-before recognized organism-like pneumonia, was found about six months later, although many questions continue to surround it and the disease that has come to be known as "Legionnaires' disease." Meanwhile, there have been — and are — other puzzling medical mysteries, some partly solved, some totally unsolved. This first article in a series of five examines one of those mysteries.

By Beverly Orndorff
Times-Dispatch Science Writer

On Friday, July 20, 1951, a previously healthy 23-year-old South Richmond man died at the Medical College of Virginia after an illness that had developed just a few days before.

His illness had features of encephalitis, or inflammation of the brain, and meningitis, or inflammation of the membranes covering the brain and spinal cord.

As it turned out, his death and the deaths of several more previously healthy young people over the next six days were to touch off a medical mystery that wouldn't be solved until 15 years later, and that still has many unanswered questions.

Here is how the story unfolded.

The 23-year-old man had been admitted to MCV on July 16, 1951, with an illness that began with an unrelenting headache, fever, loss of appetite, nausea and vomiting. Mental changes, coma and death quickly followed.

FOUR DAYS after the man died, a 10-year-old South Richmond boy, with similar symptoms, also died at MCV. The next day, a 14-year-old Glen Allen girl

died there. And a day later, a 14-year-old Chester boy died, also at MCV.

In addition, a Camp Lee officer trainee died there on July 22 with similar symptoms.

An immediate suspicion by city and area health officials was that an outbreak of a mosquito-transmitted disease, Eastern equine encephalitis, was beginning.

As a result, a control effort was initiated. Areas in South Richmond were sprayed with DDT; oil was spread on stagnant waters to reduce mosquito populations.

Someone had asked the city's health director whether it was true that all the victims had been swimming at one or the other of two Chesterfield County lakes shortly before they became ill. The director confirmed that report but said that was not necessarily significant, since encephalitis is not transmitted by water.

THERE WERE NO MORE DEATHS that summer, and the scare that was stirred, as reflected by newspapers' attention to the episode, subsided during the next few weeks.

Then in mid-August came the report of the epidemiological team from the Public Health Service that had visited Richmond to investigate circumstances surrounding the deaths.

It was not encephalitis, the report concluded, because the virus that causes encephalitis could not be recovered from any of the victims' specimens. The report left open the possibility that the disease

had been caused by some kind of bacteria, but no such bacteria were ever found.

The report did note that all the victims had been swimming in two outdoor areas before their illnesses.

The following July, the disease struck again. There were four more deaths of young, previously healthy persons in the greater Richmond area, all within a week.

THIS TIME, there was no talk of encephalitis, but local and state medical authorities had no idea what was going on. There was a brief spate of rumors about swimming areas being closed, but the city's health director denied them.

At MCV, one of those who puzzled over the cases of the summers of 1951 and 1952 was Dr. Joao G. dos Santos, who was in his residency training in clinical pathology at the time and who now is professor of clinical pathology.

As a clinical pathology resident, he had

examined the spinal fluids of those first patients in 1951 and was to become extremely interested in the disorder — whatever it was — from that point to the present.

During the ensuing years, he kept detailed files on seven of the cases from 1951 and 1952. (The family of one of the 1951 victims had refused an autopsy, and the evidence that the Camp Lee soldier really had the disorder is questionable to Dr. Santos.)

THOSE SEVEN CASES troubled him, and he often reviewed them in his mind, wondering what might have been missed.

For the next 14 years after the last 1952 case, no more episodes of the illness were recognized. (Actually, two occurred in 1957, but they weren't identified at the time.)

Then, in 1966, the case that led to the

Continued on Page 2, Col. 1

MEDICAL MYSTERIES

FIGURE 5.   A medical mystery is solved after 15 years of puzzling questions. (From *Richmond Times-Dispatch*, September 3, 1978. With permission.)

4C   St. Louis Globe-Democrat Thurs. Aug. 28, 1980

News/digest

Rare amoeba is a killer

ORLANDO, Fla.•(AP) — A rare-but-deadly disease caused by an amoeba found in Florida fresh-water lakes has claimed its fourth victim — a New York youngster who spent his vacation swimming at Walt Disney World's River Country.

The disease, amoebic meningoencephalitis, attacks the nervous system and brain, doctors say. It killed two Florida children earlier this month and appears to have been the cause of death of another youngster, a state health official said.

The latest death was that of an 11-year-old boy who visited the Orlando area during the first week of August and swam at the water attraction at Disney World, said Dr. John McGarry, director of the Orange County Health Department.

The child died after the amoeba entered his nose, went through the nasal passage and attacked the nervous system, including the brain, according to Dr. Robert Gunn, state epidemiologist.

FIGURE 6.   Another case of PAM reported from Orlando, Fla. (From *St. Louis Globe-Democrat*, August 28, 1980. With permission from the Associated Press and the *St. Louis Globe-Democrat*.)

REFERENCES

1. Alexeieff, A., Sur la division nucléaire et l'enkystemente chez quelques amibes du groupe *Limax*. I. *Amoeba punctata* Dangeard, *C. R. Seances Soc. Biol. Filiales (Paris)*, 70, 455, 1911.
2. Alexeieff, A., Sur la division nucléaire et l'enkystemente chez quelques amibes du groupe *limax*. II. *Amoeba limax* (Enend.Vahlk), *C. Seances Soc. Biol. Filiales (Paris)*, 70, 534, 1911.
3. Alexeieff, A., Sur la division nucléaire et l'enkystemente chez quelques amibes du groupe *limax*. III. *Amoeba densa* n.sp. *Amoeba cincumgranosa* n.sp. Conclusiones generales, *C. R. Seances Soc. Biol. Filiales (Paris)*, 70, 534, 1911.
4. Alexeieff, A., Sur les caractères cytologiques et la systematique des amibes du groupe *limax (Naegleria* nov. gen. et *Hartmannia* nov. gen) et des amibes parasites des vertebrates *(Protomoeba*, nov. gen), *Bull. Soc. Zool. (France)*, 37, 55, 1912.
5. Anderson, K., Jamieson, A., Jadin, J. B., and Willaert, E., Primary amoebic meningoencephalitis, *Lancet*, 1, 672, 1973.
6. Anderson, K. and Jamieson, A., Primary amoebic meningoencephalitis, *Lancet*, 1, 902, 1972.
7. Anderson, K. and Jamieson, A., Primary amoebic meningoencephalitis, *Lancet*, 2, 379, 1972.
8. Arce Vela, R. and Asato Higa, C., Encefalitis amebiana por *Acanthamoeba castellanii* (Spanish), *Diagnóstico (Lima, Perú)*, 3, 25, 1979.
9. Armstrong, J. and Pereira, M., Identification of "Ryan virus" as an amoeba of the genus *Hartmannella*, *Br. Med. J.*, 1, 212, 1967.
10. Asami, K. and Nakamura, N., Morphology and taxonomical position of an amoeba which spontaneously contaminated the agar plate of Shigella culture, *Kitasato Arch. Exp. Med. (Japanese)*, 24, 295, 1951.

11. Awerinzew, S., Über die stellung im system und die klassifizierung der protozoen, *Biol. Centralbl.*, 30, 465, 1910.
12. Boeck, W. C. and Drbohlav, J., The cultivation of *Endamoeba histolytica, Am. J. Hyg.*, 5, 371, 1925.
13. Butt, C. G., Amebic meningoencephalitis, *Am. J. Clin. Pathol.*, 42, 516, 1964.
14. Butt, C. G., Primary amebic meningoencephalitis, *N. Engl. J. Med.*, 274, 1473, 1966.
15. Butt, C. G., Baró, C., and Knorr, R. W., Pathologic progress in amebic encephalitis, *Am. J. Clin. Pathol.*, 49, 256, 1968.
16. Callicott, J. H., Jr., Amebic meningoencephalitis due to free-living amebas of the *Hartmannella (Acanthamoeba)-Naegleria)* group, *Am. J. Clin. Pathol.*, 49, 84, 1968.
17. Callicott, J. H., Nelson, E. C., Jones, M. M., dos Santos, J. G., Utz, J. P., Duma, R., and Morris, J. V., Meningoencephalitis due to pathogenic free-living amoebae, *J. A. M. A.*, 206, 579, 1968.
18. Carter, R. F., Primary amoebic meningoencephalitis: clinical, pathological and epidemiological feature of six fatal cases, *J. Pathol. Bacteriol.*, 96, 1, 1968.
19. Carter, R. F., Primary meningoencephalitis: an appraisal of present knowledge, *Trans. R. Soc. Trop. Med. Hyg.*, 66, 193, 1972.
20. Casemore, D. P., Contamination of virological tissue culture with a species of free-living soil amoeba, *J. Clin. Pathol.*, 22, 254, 1969.
21. Castellani, A., An amoeba found in cultures of a yeast: preliminary note, *J. Trop. Med. Hyg.*, 33, 160, 1930.
22. Castellani, A., An amoeba growing in cultures of a yeast: second note, *J. Trop. Med. Hyg.*, 33, 188, 1930.
23. Castellani, A., An amoeba found in cultures of a yeast: third note, *J. Trop. Med. Hyg.*, 33, 221, 1930.
24. Castellani, A., An amoeba growing in cultures of a yeast: fourth note, *J. Trop. Med. Hyg.*, 33, 237, 1930.
25. Castellani, A., Phagocytic and destructive action of *Hartmannella castellanii (Amoeba castellanii)* on pathogenic, encapsulated yeast-like fungi *(Torulopsis neoformans), Ann. Inst. Pasteur,* 89, 1, 1955.
26. Červa, L., Novak, K., and Culbertson, C. G., An outbreak of amoebic meningoencephalitis, *Am. J. Epidemiol.*, 88, 436, 1968.
27. Chang, R. S., Goldhaber, P., and Dunnebacke, T. H., The continuous multiplication of lipovirus-infected human cells, *Proc. Nat. Acad. Sci. U.S.A.*, 52, 709, 1964.
28. Chang, R. S., Pan, I. H., and Rosenau, B. J., On the nature of the "Lipovirus", *J. Exp. Med.*, 124, 1153, 1966.
29. Chatton, E., Sur quelques genres d'amibes libres et parasites. Synonymies, homonymie, impropriété, *Bull. Soc. Zool. (France)*, 37, 109, 1912.
30. Chatton, E. et Lalung-Bonnaire, E., *Amibe limax (Vahlkampfia* n. gen.) dans l'intestin humain. Son importance pour l'interprétation des amibes de culture, *Bull. Soc. Pathol. Exotique (Paris)*, 5, 135, 1912.
31. Chi, L., Vogel, J. E., and Shelokov, A., Selective phagocytosis of nucleated erythrocytes by cytotoxic amebae in cell culture, *Science*, 130, 1763, 1959.
32. Culbertson, C. G., *Acanthamoeba:* observations on animal pathogenicity, *Science*, 127, 1506, 1958.
33. Culbertson, C. G., Pathogenic *Acanthamoeba (Hartmannella), Am. J. Clin. Pathol.*, 35, 195, 1961.
34. Culbertson, C. G., Ensminger, P. W., and Overton, W. M., Pathogenic *Hartmannella (Acanthamoeba)* (abstr.), *Am. J. Clin. Pathol.*, 42, 528, 1964.
35. Culbertson, C. G., Ensminger, P. W., and Overton, W. P., Pathogenic *Naegleria* sp. Study of a strain isolated from human cerebrospinal fluid, *J. Protozool.*, 15, 353, 1968.
36. Cursons, R. T. M. and Brown, T. J., The 1968 cases of primary amoebic meningoencephalitis — *Myxomycete* or *Naegleria?*, *N.Z. Med. J.*, 82, 123, 1975.
37. Cursons, R. T. M., Brown, T. J., and Culbertson, C. G., Immunoperoxidase staining of trophozoites in primary amoebic meningoencephalitis, *Lancet*, 2, 479, 1976.
38. Dangeard, P. A., Étude de la Karyokinèse chez l' *Amoeba hyalina* sp. nova. *Botaniste*, 7, 49, 1900.
39. Das, S. R., A simple and reliable method for obtaining clones of *Entamoeba histolytica* and other amoebae from cultures, *Curr. Sci.*, 41, 736, 1972.
40. de la Arena, J. F., *Astramoeba torrei* N. sp. de amiba de agua dulce, *Mem. Soc. Cubana Hist. Nat.*, 21, 77, 1952.
41. de la Arena, J. F., Nueva especie de amiba del género mayorella Schaeffer, *Mem. Soc. Cubana Hist. Nat.*, 21, 315, 1953.
42. de la Arena, J. F., Nueva especie de amiba de Cuba, *Mem. Soc. Cubana Hist. Nat.*, 22, 15, 1955.
43. Derrick, E., A fatal case of generalized amoebiasis due to protozoan closely resembling if not identical with *Iodoamoeba butschlii, Trans. R. Soc. Trop. Med. Hyg.*, 42, 191, 1948.
44. Diamond, L. S., Techniques of axenic cultivation of *Entamoeba histolytica* Schaudinn, 1903 and *E. histolytica* like amebae, *J. Parasitol.*, 54, 1047, 1968.
45. dos Santos, J. G., Fatal primary amebic meningoencephalitis: a retrospective study in Richmond, Virginia, *Am. J. Clin. Pathol.*, 54, 737, 1970.

46. Douglas, M., Notes on the classification of the amoeba found by Castellani in cultures of a yeast-like fungus, *J. Trop. Med. Hyg.,* 33, 258, 1930.

47. Dujardin, F., Historie naturelle des zoophytes. Infusoires, *Libraire Encyclopedique de Roret (Paris),* 1841.

48. Duma, R. J., Farrell, W. H., Nelson, E. C., and Jones, M. M., Primary amebic meningoencephalitis, *N. Engl. J. Med.,* 24, 1315, 1969.

49. Dunnebacke, T. H. and Williams, R. C., A reinterpretation of the nature of "lipovirus" cytopathogenicity, *Proc. Natl. Acad. Sci. U.S.A.,* 57, 1361, 1967.

50. Ehrenberg, C. G., Die infusionsthierchen als vollkommene organismen, Leipzig, 1838.

51. Eldridge, A. and Tobin, J., Ryan virus, *Br. Med. J.,* 1, 299, 1967.

52. Flexner, S., Amoebae in an abscess of the jaw, *Johns Hopkins Hosp. Bull.,* 3, 104, 1892.

53. Fowler, M. and Carter, R. F., Acute pyogenic meningitis probably due to *Acanthamoeba* spp.: a preliminary report, *Br. Med. J.,* 2, 740, 1965.

54. Gros, G., Fragments d'helminthologie et de physiologie microscopique, *Bull. Soc. Imp. Nat. (Moscow),* 22, 549, 1849.

55. Hartmann, M., Untersuchungen uber parasitische Amoeben, *Arch. Protistenk,* 18, 207, 1910.

56. Herrera, J. M., Meningoencefalitis por ameba histolytica, *Arch. Méd. Panameños (Spanish),* 1, 149, 1952.

57. Hewitt, R., The natural habitat and distribution of *Hartmannella castellanii* (Douglas). A reported contaminant of bacterial cultures, *J. Parasitol.,* 23, 491, 1937.

58. Hogue, M. J., Studies in the life history of an amoeba of the limax group *Vahlkampfia calkinsi* n.sp., *Arch. Protistenk,* 35, 154, 1914.

59. Hogue, M. J., Studies on the life history of *Vahlkampfia patuxent* n. sp, parasitic in the oyster with experiments regarding its pathogenicity, *Am. J. Hyg.,* 1, 321, 1921.

60. Hollande, A., Etude cytologique et biologique de quelques flagelles libres, *Arch. Zool. Exp. Gen.,* 83, 1, 1942.

61. Jadin, J. B. and Willaert, E., Au sujet des meningites amibiennes, *Piscine,* 34, 54, 1972.

62. Jadin, J. B., De la meningoencephalite amibienne et du pouvoir pathogene des amibes "limax", *Ann. Biol. (Paris),* 12, 305, 1973.

63. Jager, B. V. and Stamm, W. P., Brain abscesses caused by free-living amoeba probably of the genus *Hartmannella* in a patient with Hodgkin's disease, *Lancet,* 2, 1343, 1972.

64. Jahnes, W. G., Fullmer, H. M., and Li, C. P., Free-living amoebae as contaminants in monkey kidney tissue cultures, *Proc. Soc. Exp. Biol. Med.,* 96, 484, 1957.

65. Johnson, C. M., Un caso de meningoencefalitis primaria amibiana, *Rev. Méd. Panamá (Spanish),* 2, 141, 1977.

66. Kenney, M., The micro-Kolmer complement fixation test in routine screening for soil ameba infection, *Health Lab. Sci.,* 8, 5, 1971.

67. Kernohan, J., Magath, T. B., and Schloss, G. T., Granuloma of brain probably due to *Endolimax williamsi (Iodamoeba butschlii),* *Arch. Pathol.,* 70, 576, 1960.

68. Leidy, J., Freshwater rhizopodes of North America, *Rep. U.S. Geol. Surv. Territories,* 12, 1, 1879.

69. Ma, P., Willaert, E., Juechter, K. B., and Stevens, A. R., A case of keratitis due to *Acanthamoeba* in New York, New York, and features of 10 cases, *J. Infect. Dis.,* 143, 662, 1981.

70. Mandal, B. N., Gudex, D. J., Fitchett, M. R., Pullon, D. H., Maldoch, J. A., David, C. M., and Apthorp, J., Acute meningoencephalitis due to amoeba of the order *Myxomycetale* (lime mould), *N.Z. Med. J.,* 71, 16, 1970.

71. Martinez, A. J., dos Santos, J. G., Nelson, E. C., Stamm, W. P., and Willaert, E., Primary amebic meningoencephalitis, in *Pathology Annual,* Vol. 12, Sommers, S. C. and Rose, P. P., Eds., Appleton-Century-Crofts, New York, 1977, 225.

72. Martinez, A. J., Is *Acanthamoeba* encephalitis an opportunistic infection?, *Neurology,* 30, 567, 1980.

73. Moore, A. E. and Hlinka, J., *Hartmannella* sp. *(Acanthamoeba)* as a tissue culture contaminant, *J. Natl. Cancer Inst.,* 40, 569, 1968.

74. Nägler, K., Entwicklungsschichtliche studien über amöben, *Arch. Protistenk,* 15, 1, 1909.

75. Nakamura, N., On a strain of amoeba accidentally discovered on agar medium which is phagocytic to bacteria, *Kitasato Arch. Exp. Med.,* 24, 23, 1951.

76. Nakanishi, K., A new type of ameba *(Amoeba ferox)* phagocyting pathogenic intestinal bacteria, recovered from river water in Java, *Jpn. Med. J.,* 3, 231, 1950.

77. Nelson, E. C. and Jones, M., Culture isolation of agents of primary amebic meningoencephalitis, *J. Parasitol.,* 56, 248, 1970.

78. Nelson, E. C., Procedures for the isolation of pathogenic *Naegleria* (abstr.), *Va. J. Sci.,* 23, 145, 1972.

79. Page, F. C., Taxonomic criteria for *Limax* amoebae with descriptions of 3 new species of *Hartmannella* and 3 of *Vahlkampfia, J. Protozool.,* 14, 499, 1967.

80. Patras, D. and Andujar, J. J., Meningoencephalitis due to *Hartmannella (Acanthamoeba)* (abstr.), *Am. J. Clin. Pathol.,* 44, 580, 1965.

81. Patras, D. and Andujar, J. J., Meningoencephalitis due to *Hartmannella (Acanthamoeba)*, *Am. J. Clin. Pathol.*, 46, 226, 1966.

82. Pereira, M., Marsden, H., Corbitt, G., and Tobin, J., Ryan virus: a possible new human pathogen, *Br. Med. J.*, 1, 130, 1966.

83. Prowazek, S., Weitere Beitrag zur Kenntnis der Entamoeben, *Arch. Protistenk*, 26, 241, 1912.

84. Rafalko, J., Cytological observation on the amoebo-flagellate *Naegleria gruberi*, *J. Morphol.*, 81, 1, 1947.

85. Robert, V. B. and Rorke, L. B., Primary amebic encephalitis probably from *Acanthamoeba*, *Ann. Intern. Med.*, 79, 174, 1973.

86. Rondanelli, E. G., Carosi, G., Minoli, L., and Filice, G., Le meningoencefaliti amebiche primarie (MAP) da amebe del grupo *Hartmannella-Naegleria*. Un capitolo nuovo de patologia amebica, *Terapia*, 57, 136, 1972.

87. Rondanelli, E. G., Carosi, G., Filice, G., and Scaglia, M., Attualita in tema di amebiasis: quadri ultrastrutturali e nosografia dell'amebiasis da *E. histolytica* e delle meningoencefaliti da amebe "a vita libera" patogene per l'uomo, in *Basi Biologiche Medicina Moderna*, Med Scientifice, Torino, 3, 305, 1980.

88. Rösel von Rosenhof, A. J., Der kleine proteus. Der monat.-herausgeg, *Insekten-Belustigungen*, 3, 622, 1755.

89. Schaeffer, A., Taxonomy of the Amoebas with Descriptions of 39 New Marine and Freshwater Species, Vol. 24, Publ. 345, Carnegie Institute of Washington, Washington, D.C., 1926, 1.

90. Schardinger, F., Der entwicklungskreis einer *Amoeba lobosa (Gymnamoeba): Amoeba gruberi*, Sitzb Kaiserl Akad. Wiss., Vienna, 1, 108, 713, 1899.

91. Shinn, L. E. and Hadley, P. B., Note on the spontaneous contamination of a bacterial culture by an organism resembling *Hartmannella castellani*, *J. Infect. Dis.*, 58, 23, 1936.

92. Shookhoff, H. B., (Editorial) Meningoencephalitis due to free-living amoebas normally found in soil, *Ann. Intern. Med.*, 70, 1276, 1969.

93. Shumaker, J. B., News. Primary amebic meningoencephalitis, *J. Infect. Dis.*, 121, 89, 1970.

94. Singh, B. N., Selectivity in bacterial food by soil amoeba in pure mixed culture and in sterilized soil, *Ann. Appl. Biol.*, 28, 52, 1941.

95. Singh, B. N., A culture method for growing small free-living ameba for the study of their nuclear division, *Nature*, 165, 65, 1950.

96. Singh, B. N., Nuclear division in nine species of small free-living amoebae and its bearing on the classification of the order *Amoebida*, *Philos. Trans. R. Soc. London B*, 236, 405, 1952.

97. Singh, B. N., Selectivity in bacterial food by soil amoebae in pure mixed culture and in sterilized soil, *Ann. Appl. Biol.*, 28, 52, 1964.

98. Symmers, W.St. C., Primary amoebic meningoencephalitis in Britain, *Br. Med. J.*, 4, 449, 1969.

99. Vahlkampf, E., Beiträge zur biologie und entwick lungsgeschichte von *Amoeba limax* einsch lie blich der Züchtung auf Künstlichen Nährböden, *Arch. Protistenk*, 5, 167, 1905.

100. Volkonsky, M., *Hartmannella castellanii* Douglas, et classification des hartmannelles, *Arch. Zool. Exp. Gene*, 72, 317, 1931.

101. Wang, S. S. and Feldman, H. A., Occurrence of *Acanthamoeba* in tissue cultures inoculated with human pharyngeal swabs, *Antimicrob. Agents Chemother.*, 1, 50, 1961.

102. Wang, S. S. and Feldman, H. A., Isolation of *Hartmannella* species from human throats, *N. Engl. J. Med.*, 277, 1174, 1967.

103. Warhurst, D. and Armstrong, J., A study of a small amoeba from mammalian cell cultures infected with Ryan virus, *J. Gen. Microbiol.*, 50, 207, 1968.

104. Zulueta, A. de, Promitosis y sindiéresis; dos modos de división nuclear coexistentes en amebas del grupo "limax", *Trabajos Mus. Nac. Cien. Nat. Ser. Zool. (Madrid)*, 33, 1, 1917.

Chapter 3

PROTOZOOLOGY, TAXONOMY, AND NOMENCLATURE OF FREE-LIVING AMEBAS

I. CLASSIFICATION OF FREE-LIVING AMEBAS: AN OVERVIEW

Several classifications of free-living amebas based mainly on morphological features have been proposed and new species of high and low virulence described.[21,30,139,153-155] However, since the morphological features alone are tenuous and common to many free-living or amphizoic amebas, they do not allow clear criteria for definite classification.[2,16,18,54,125,156] It is necessary to supplement morphological characteristics with other parameters like antigenic properties, growth, and nutritional characteristics, solubility and patterns of hydrosoluble proteins, biochemical composition, mitosis, motility, and virulence.[23,34,36] A new approach that might improve classification within the genus and complement morphological, immunological, and isoenzyme patterns is the use of mitochondrial DNA instead of the measurement of the nuclear DNA.[15] Singh believes that the principal features upon which classification should be based are nuclear structures and patterns of division during mitosis.[127-131] Dr. Singh recognized both the genera *Acanthamoeba* and *Hartmannella* on locomotive form and behavior, the former producing acanthopodia and the latter having a limax form. Since the amebas included in these genera divide by mesomitosis, he placed them in the family Hartmannellidae and the genus *Hartmannella*. The amebas causing meningoencephalitis in humans and lower mammals belong to *Acanthamoeba* and not *Hartmannella*. These two genera are also distinct serologically and nutritionally.[130]

Dr. Singh studied the antigenic relationship among *Acanthamoeba rhysodes, A. culbertsoni, A. polyphage, A. palestinensis, A. invadens, A. glebae,* and *A. astronyxis.* He found that all the species share some common soluble antigens, suggesting that they are characteristic of the genus *Acanthamoeba. A. glebae* and *A. invadens* have smooth, rounded cysts without any pores or opercula. Thus, *Acanthamoeba* should be recognized on locomotive behavior and not on cystic character.[131] Page, on the other hand, proposed a classification based on locomotion and motility with pseudopodial and cytological features, nuclear structures, division pattern during mitosis, ultrastructure, nutrition, and immunological characteristics. Dr. Page recognizes *Acanthamoeba* as distinct from *Hartmannella,* based on such features as acanthopodia, locomotion, and the morphology of cysts. This classification does not consider pathogenicity as a distinguishing character; therefore, it considers all amebas as potentially pathogenic, either to man or to animals[74-84] (Tables 1 and 2).

The classification of free-living amebas by Chang represents a combination of the ideas of both Singh and Page and is based on phylogenetic principles.[26]

Pussard,[91-99] Molet and Kremer, and Molet and Ermolieff-Braun[68,69] emphasized that important features for taxonomic classification of free-living amebas should be based on: nuclear patterns during mitosis, the appearance of centrospheres, morphological characteristics of the cyst wall, flagellation tests, and virulence or results of the animal inoculation.

The human pathogens known to date would be classified accordingly: *Naegleria* under the family Vahlkampfiidae, *Hartmannella* under the family Hartmannellidae, *Acanthamoeba* under the family Acanthopodinae, and *Entamoeba* under the family Entamoebidae (Tables 1 and 2).

Table 1
PROPOSED CLASSIFICATION OF NAKED AMEBAS ACCORDING TO PAGE[83]

Phylum Sarcomastigophora
Subphylum Sarcodina
Superclass Rhizopodia
Class Lobosea
Subclass Gymnamoebia
Order Amoebida
Suborder Tublina
Family Amoebidae
Family Hartmannellidae
Family Entamoebidae[a]
Suborder Conopodina
Family Paramoebidae[b]
Suborder Acanthopodina
Family Acanthamoebidae[c]
Family Echinomoebidae
Suborder Flabellina
Family Flabellulidae
Family Hyalodiscidae
Suborder Thecina
Family Thecamoebidae
Order Schizopyrenida
Family Vahlkampfiidae
Order Pelobiontida
Family Pelomyxidae

[a]   *E. histolytica* in the family Entamoebidae is not considered a "free-living" ameba.
[b]   More recently, Page has suggested that the Acanthamoebae be placed under the family Paramoebidae.
[c]   *Acanthamoeba* are considered as *Hartmanella* by Singh.

II. MORPHOLOGICAL CHARACTERISTICS OF *NAEGLERIA FOWLERI*

A. The Trophozoite (Proliferative Form)

The light microscopic features of trophozoites and cysts of free-living amebas should be supplemented with transmission and scanning electron microscopy.[111] Phase contrast microscopy and the Nomarski optics will add more details that may help in the identification and precise classification of these amebas. *N. fowleri*, *N. aerobia*, and *N. invadens* are synonyms. *N. jadini*, *N. gruberi*, *N. lovaniensis*, and *N. australiensis* are strains of *Naegleria*, that morphologically are indistinguishable from one another, but they differ in virulence, pathogenicity, and antigenically (by immunoelectrophoresis and agglutination techniques).[6,7,17,19,33,37-42] *N. australiensis* is mildly pathogenic for mice and characterized by longer incubation period and lower mortality rate than *N. fowleri*. *N. gruberi* is not pathogenic.

The amebas which have been isolated from man are morphologically identical to the common, free-living *N. gruberi*.[22,28,29,48] The trophozoites or the proliferative form of *N. fowleri* are active and variable in size and shape, and characterized by round or blunt pseudopodia called lobopodia (Figure 1a and 1b). When rounded they measure from about 8 to 30 μm in diameter. Broadly rounded, granule-free processes erupt from the surface and granular cytoplasm flows into them. When streaming continues in one direction the cell becomes finger-like in shape. More often, however, protrusions are formed in quick succession at different points on the surface of the cell so that its shape is constantly changing. The cytoplasm is finely granular and contains a conspicuous, clear nucleus and dense central nucleolus giving it the appearance of a halo. Water, contractile, and food vacuoles are usually apparent. The water vacuole

Table 2
CRITERIA FOR ORDERS AND FAMILIES CONTAINING PATHOGENIC AMEBAS, ACCORDING TO PAGE'S PROPOSED CLASSIFICATION OF NAKED AMEBAS, 1976[83]

Order Amoebida (Kent, 1880) — typically uninucleate; mitochondria typically present; no true bidirectional flow of cytoplasm; no flagellate stages known
   Suborder Tubulina (Bovee and Jahn, 1966, emend) — body a branched or unbranched cylinder, with hyaline cap present or absent; nuclear division mesomitotic
      Family Hartmannellidae (Volkonsky, 1931; emend. Page, 1974) — monopodial (limax) amebas with locomotion by generally steady flow, sometimes with gentle hemispherical bulging to either side at anterior end; with vesicular nucleus; cysts, if formed, uninucleate, in one genus often binucleate
      Family Entamoebidae (Chatton, 1925) — monopodial amebas with endosome a small granule or cluster of granules; mature cyst, if cyst occurs, usually with 4, 8, or more nuclei produced by divisions within cyst; all known species except one parasitic
   Suborder Acanthopodina — more or less finely tipped, sometimes filose, often furcate hyaline subpseudopodia produced from a broader hyaline zone; not regularly discoid, though sometimes expanded on substratum; usually forming cysts; nuclear division mesomitotic or metamitotic
      Family Acanthamoebidae (Sawyer and Griffin, 1975) — several to many subpseudopodia (acanthopodia), usually pointed but sometimes blunt; outline of ameba in locomotion oval, more or less triangular, elongate, or irregular; extranuclear centrospheres reported; cysts polyhedral or thickly biconvex, with cellulose-containing wall consisting of more or less polygonal or stellate endocyst and more or less rippled ectocyst; excystment by removal of operculum at point of contact between endocyst and ectocyst
Order Schizopyrenida (Singh, 1952) — body a monopodial cylinder (limaciform), usually moving with more or less strongly eruptive, hyaline, hemispherical bulges; typically uninucleate; nuclear division promitotic; temporary flagellate stages common
   Family Vahlkampfiidae (Jollos, 1917; Zulueta, 1917) — with the character of the order. (Vahlkampfia, Naegleria, Adelphamoeba, Tetramitus, Heteramoeba.)

Note: See also Table 1.

bursts at intervals and is reformed by the confluence of smaller vacuoles. Virus-like particles within trophozoites have been implicated in pathogenicity.[48,49,11,119] When trophozoites of *N. fowleri* are placed in distilled water they may develop two or more flagella (Figure 2a and 2b). The flagellated state is temporary and reverses to the ameboid form after about 24 hr.[45-47,87,143] The outer surface of trophozoites is irregular (Figure 3) and points of contact between them are conspicuous. Sucker-like structures, or amebastome, have been demonstrated (Figure 4) as determinants of virulence, invasiveness, and phagocytic capabilities of *N. fowleri* (Figure 5).[60]

Well-preserved *Naegleria* are bound by unit membranes. Each membrane is composed of two electron-dense layers separated by an electron-transparent zone. Throughout the cytoplasm a poorly organized, rough endoplasmic reticulum is seen, made up of elongated tubular vesicles covered with ribosomes and not closely associated to mitochondria. Most impressive is the presence of many dumbbell-shaped mitochondria. These are bound by two membranes separated by a clear space, the outer membrane being thinner than the inner one. The cristae are thick, prominent, and irregularly arranged. Cup-shaped, spherical, and oval mitochondria are also found. These contained tubular, transverse, and oblique cristae (Figure 6).

Three types of vacuoles are noted. Most numerous are empty vacuoles found within the cytoplasm. The next most frequent vacuole is bound by a unit membrane and contains electron-dense material which is often surrounded by whorl-like formations. These formations may possibly have represented lysosomes. Some of the larger ones are thought to be food vacuoles, since they contained "engulfed" material and cellular debris, while others contained "myelin figures". When isolated from experimentally infected animals, many *Naegleria* contain ingested red blood cells and leukocytes within within their cytoplasm. Small, electron-dense granules of glycogen are scattered throughout the cytoplasm, as well as phospholipid droplets.

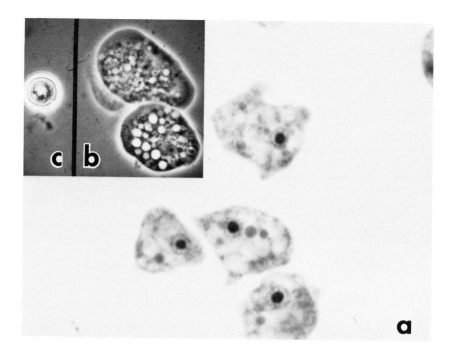

FIGURE 1.   (a) Trophozoites of *Naegleria fowleri* (Lee strain) demonstrating the blunt lo-comotor pseudopodium or lobopodium. The cytoplasm is abundant with multiple organelles. The characteristic nucleus contains a centrally placed dense nucleolus. (Iron-hematoxylin; magnification × 800.) (b) Trophozoites with Nomarski optics. (Magnification × 600.) (c) Cyst of *N. fowleri* (Lee strain) with Nomarski technique. (Magnification × 600.)

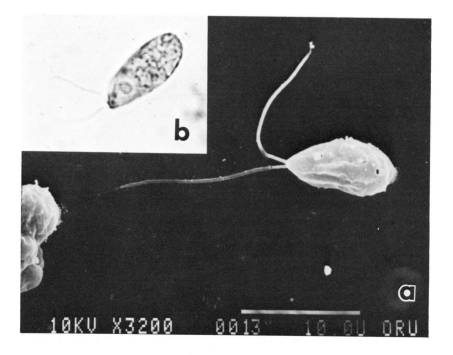

FIGURE 2.   (a) Flagellated trophozoites of *Naegleria fowleri* from axenic culture. (Magni-fication × 1000.) (b) Scanning electron micrograph. (Magnification × 5000.) (Supplied by Drs. David T. John and Thomas B. Cole, Jr., of Oral Roberts University.)

25

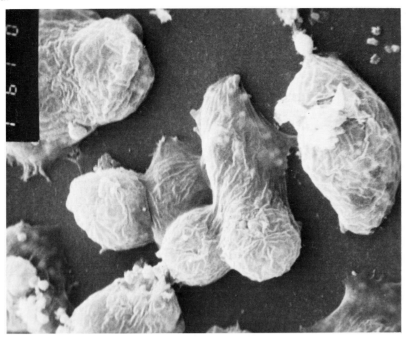

FIGURE 3.   Scanning electron micrograph of trophozoites of *Naegleria fowleri* (CJ strain) growing in axenic medium. The cell membrane is wrinkled. Spherical cytoplasmic organelles may be seen within the cytoplasm. Points of contact are evident. (Magnification × 3000.)

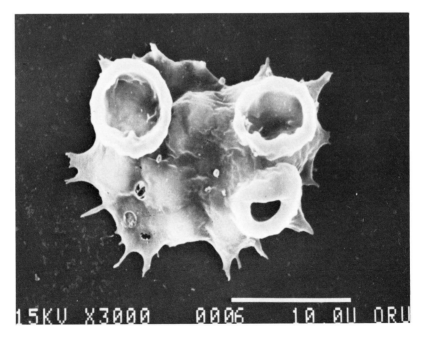

FIGURE 4.   Trophozoite of *Naegleria fowleri* (CJ Strain) from axenic culture showing three smooth-edged suckers (amebastomes). Scanning electron micrograph; bar — 5 μm. (Courtesy of Drs. David T. John and Thomas B. Cole, Jr., of Oral Roberts University.)

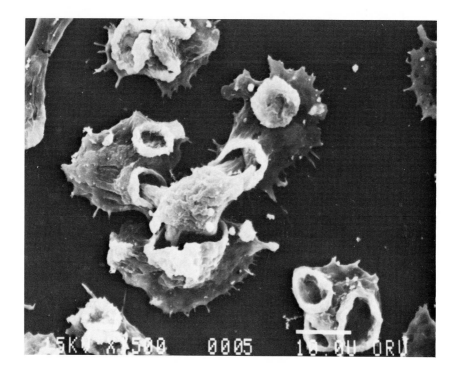

FIGURE 5.   Trophozoites of *Naegleria fowleri* (CJ Strain) from axenic culture using amebastomes to attack and engulf portions of a presumably dead amebic trophozoite. Scanning
electron micrograph; bar — 10 μm. (From John, D. T., Cole, T. B., and Marciano-Cabral, F.
M., *Appl. Env. Microbiol.*, 47(1), 12, 1984. With permission.)

The nucleus is bound by two, clearly visible, electron-dense membranes, separated
by a clear space. A few pores can be seen within the nuclear envelope. In the central
portion of the nucleus is a dense, prominent nucleolus or endosome (karyosome). In
these amebas, no Golgi apparatus is clearly recognized. The locomotor pseudopodia
are composed of finely granular, electron-dense material, distinct from the rest of the
cytoplasm. These pseudopodia are free of the vacuoles and particles that fill the rest
of the cytoplasm.

B. The Cysts

The cysts are very useful in classification of free-living amebas because there is no
distinctive or specific features to differentiate genera and species. The most characteristic features of the cysts are the shape, the appearance of the wall, and the number
and morphology of the pores.[65,120]

The cysts of *N. gruberi* and *N. australiensis* are spherical with a smooth, single-
layered wall (Figure 1c) and have one or two minute pores which appear to be plugged
and like craters of volcanoes (Figure 7). The pores of *N. fowleri* are flat. The cytoplasm
of the cyst is finely granular and the characteristic nucleus lies centrally. Cellulose is a
known component of the cyst wall in *Acanthamoeba*.[137] Chitin is also present within
the cyst wall of *Entamoeba invadens*.[135,138]

III. MORPHOLOGICAL CHARACTERISTICS OF *ACANTHAMOEBA* SPP.

A. The Trophozoite (Proliferative From)

The better known species of *Acanthamoeba* spp. are *A. castellanii, A. astronyxis,*

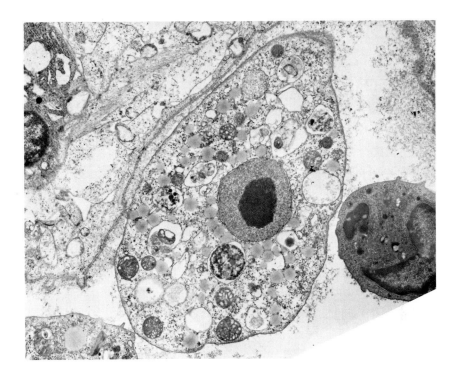

FIGURE 6.   *Naegleria fowleri* (Lee strain) growing in mouse brain. The cytoplasm contains numerous organelles. The granular nuclear chromatin contains a centrally placed dense nucleolus. (Magnification × 5000.)

A. culbertsoni, A. comandoni, A. polyphaga, A. terricola, Hartmannella vermiformis, A. royreba, A. palestinensis, H. exudans, H. glebae, H. rhysodes, A. mauritaniensis, A. lenticulata, and *A. griffini.* The last one is a marine ameba isolated by T. K. Sawyer in 1971.

The trophozoites of these species are easily recognized and differentiated from *Naegleria* spp. by the presence of slender, spine-like processes (Figure 8b). The trophozoites measure from 20 to 40 μm.[1,8,10-15,70,88,106-110,114,116,142,151] After being transferred to a slide, the cells become flat and assume various shapes (Figure 8a). Locomotion is by a slow, gliding movement; the edge of the cell appears fringed due to the fine hyaline processes or spikes which protrude from the surface (Figure 9). The cytoplasm is finely granular and contains a single nucleus which, like that of *Naegleria,* has a large, dense, central nucleolus surrounded by a clear zone. Cytoplasmic organelles like water, food vacuoles, and mitochondria are usually evident in the cytoplasm. Trophozoites may be identified within the CNS tissue of the infected animal or human by the spherical, centrally placed nucleolus surrounded by a clear nuclear halo (Figure 10a).

Acanthamoeba spp. amebas are found by a thin, electron-dense cytoplasmic membrane. The acanthopodia are composed of finely granular particulate matter and glycogen granules, devoid of cytoplasmic organelles and clearly distinct from the rest of the cytoplasm.[140]

The cytoplasm contains free ribosomes, some small vesicles, tubules, groups of fibrils which sometimes appear in bundles, and a variable number of cytoplasmic membrane systems or organelles such as Golgi complex, smooth and rough endoplasmic reticulum, digestive vacuoles, water expulsion vesicle or contractile vacuole, mitochondria, lipid droplets, lysosomes, and glycogen particles. This is the contractile vacuole or water expulsion vesicle which usually has a "fuzzy" border. The large food vacuoles

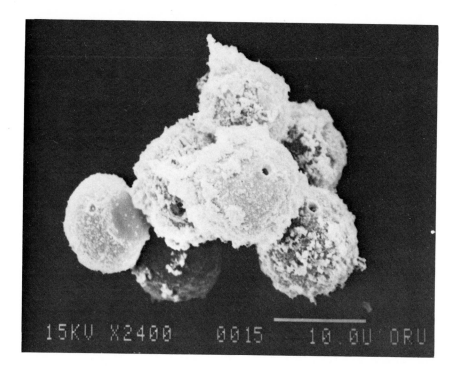

FIGURE 7.   Cysts of *Naegleria australiensis* showing the characteristic spherical shape with the crater-like pores. Scanning micrograph; bar — 10 μm. (Courtesy of Drs. David T. John and Thomas B. Cole, Jr., of Oral Roberts University.)

or digestive vesicles are characterized by typical membrane invaginations which contain numerous "myelin figures", flocculent precipitate, whorl-like structures, and different types of crystalloid and "engulfed" material. Other types of vacuoles are filled with a finely granular, sparse, electron-dense particulate matter (Figure 11).

Different forms of mitochondria can be found: spherical, circular, rod-shaped, oval, or elongated and slightly dumbbell shaped. The cristae are formed by branching, anastomosing tubular extensions of the inner mitochondrial membrane. Amorphous, electron-dense granules sometimes can be observed within mitochondria, and at other times dense intracrystalline droplets can be seen. Usually, the mitochondria measure 0.5 to 1.5 μm in length by 0.1 to 1 μm in diameter. Endosymbionts may be found in the cytoplasm.[90]

The nucleus is usually centrally located, relatively small, with a centrally placed nucleolus. An interesting feature is the sharp separation between the nuclear membrane and the cytoplasm. The mean diameter of a nucleus of a trophozoite is 6 μm.

The nuclear membrane is made up of two electron-dense membranes, separated by a clear space. At regular intervals, fusion points or electron-dense areas connect the two nuclear membranes (Figure 11).

The chromatin of the nucleus is irregularly clumped. The nucleolus is about 2.4 μm in diameter (endosome or karyosome) and is usually dense, compact, and spherical, but sometimes variable in morphology. Sometimes vacuoles or vesicles can be seen within the nucleolus.

B. The Cysts

The cysts of *Acanthamoeba* (*A. castellanii* and *A. polyphaga)* are characterized by a double walled envelope (Figure 8c). The average diameter of a cyst of *A. castellanii*

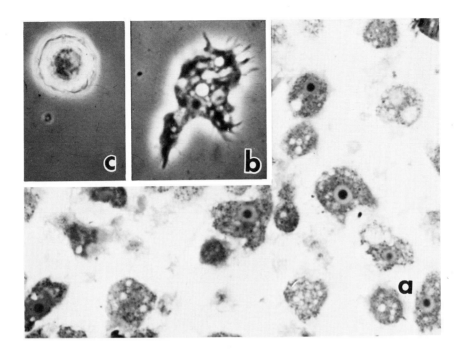

FIGURE 8. (a) Trophozoites of *Acanthamoeba glebae* showing the dense, round nucleolus encompassed by a clear nuclear halo and abundant granular cytoplasm. Culture embedded in Epon, 1-μm section. (Toluidine blue; magnification × 600.) (b) Trophozoite of *Acanthamoeba* spp. showing the slender acanthopodia. The karyosome or nucleolus is dense and centrally placed within a spherical clear nucleus. (c) Cyst of *Acanthamoeba* spp. with a thick, undulated wall. Nomarski optics. (Magnification × 800.)

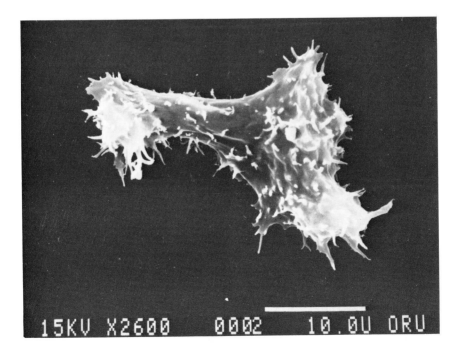

FIGURE 9. Trophozoite of pathogenic *Acanthamoeba castellanii* isolated from the environment. Scanning electron micrograph; bar — 10 μm. (Courtesy of Drs. David T. John and Thomas B. Cole, Jr., of Oral Roberts University.)

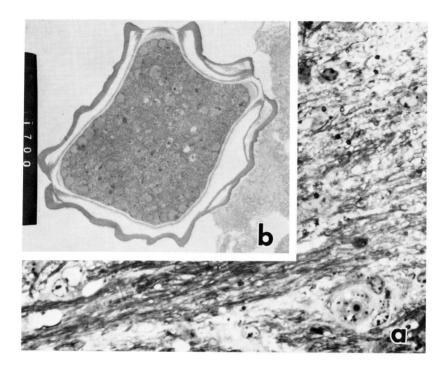

FIGURE 10.   (a) Trophozoite of *Acanthamoeba glebae* within mouse brain 15 days after intranasal inoculation. There is minimal inflammatory reaction. One micron plastic embedded section. Toluidene blue. (Magnification × 600.) (b) Cyst of *Acanthamoeba castellanii* showing a double-wrinkled wall and three ostioles. The cytoplasm is packed with numerous organelles. (Magnification × 5000.)

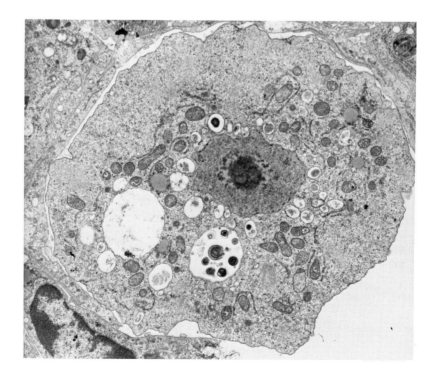

FIGURE 11.   Trophozoite of *Acanthamoeba castellanii* in mouse brain 21 days after intranasal inoculation. (Magnification × 5000.)

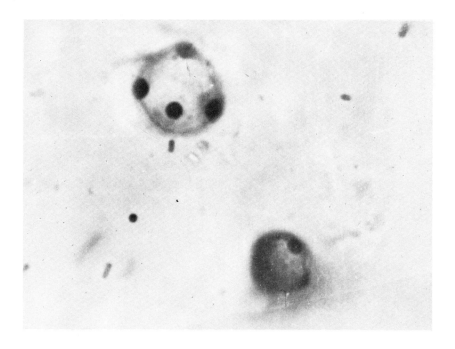

FIGURE 12.   Cysts of *Acanthamoeba polyphaga* demonstrating the pores arranged in the equatorial plane stained with the silver method of Professor Dimas Fernández-Galiano. (Magnification × 800.) (Photomicrograph courtesy of Drs. María Teresa Fernández-Galiano and Susana Serrano.)

is 16.2 μm, while the cyst of *A. polyphaga* is 12.5 μm. Different states in the encystment process can be found: spherical, "star shaped", hexagonal, polygonal with one, two, or more opercula, etc.[50,63,135] *A. palestinensis* cysts show similar characteristics.[63]

The cyst wall generally is composed of two major layers: a fibrous, wrinkled exocyst within a varying amount of matrix material showing distinct parallel layering, and the polyhedral or stellate layer which is composed of fine fibrils and forms an operculum over the ostioles. There is a space between the two layers, except at the level of the operculum in the center of the ostioles. Cytoplasmic debris derived from autolysosomes can be seen in the wall.

The cytoplasm of the cyst is very dense, packed with lipid droplets, spherical mitochondria, and lysosomes, and the cytoplasm forms bulges at each ostiole. The mitochondria may disclose electron-dense irregular structures within their cristae.

The outer wall exocyst is wrinkled, while the inner endocyst is smooth and polygonal (Figure 10b). The angles of the inner wall make contact with the outer at a number of points to form pores or ostioles. The enclosed cell takes the shape of the inner wall, its cytoplasm is finely granular, and the single nucleus is centrally placed. Cysts of *Acanthamoeba* spp. possess the pores arranged in an equatorial plane (Figure 12). Pathogenic *Acanthamoeba* spp. may show also depressed pores or ostioles (Figure 13). There may be an accumulation of Golgi vesicles and stalks of endoplasmic reticulum near the plasma membrane arranged in parallel fashion (Figure 14).

IV. COMPARISON BETWEEN *NAEGLERIA FOWLERI* AND *ACANTHAMOEBA* SPP.

A. Life Cycle of *Naegleria fowleri* (Figure 15)

The trophozoite or vegetative form of *N. fowleri* may develop from two to nine

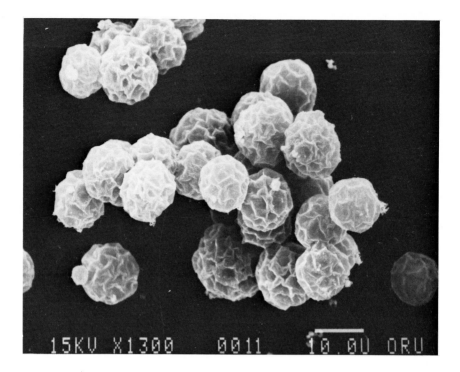

FIGURE 13.   Pathogenic *Acanthamoeba castellanii* cysts. Bar — 10 μm. (Courtesy of Drs. David T. John and Thomas B. Cole, Jr., of Oral Roberts University.)

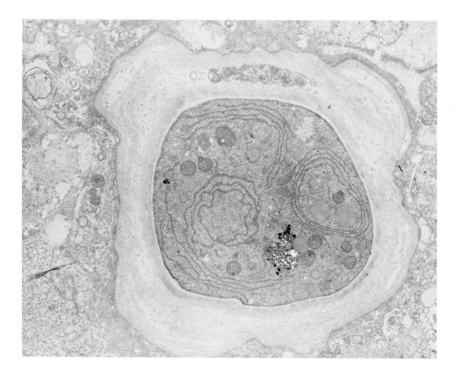

FIGURE 14.   Micrograph of a cyst of *Acanthamoeba* spp. in the lung of a mouse inoculated intranasally 2 weeks previously. (Magnification × 5000.)

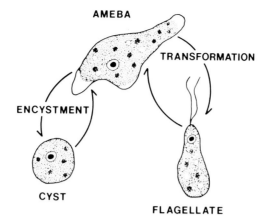

FIGURE 15. Life cycle of *Naegleria fowleri.* (Courtesy of Dr. David T. John and Dr. Thomas B. Cole, Jr., of Oral Roberts University.)

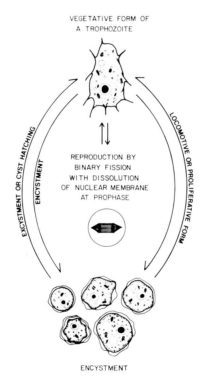

FIGURE 16. Life cycle of *Acanthamoeba* spp.

flagella temporarily when in contact with distilled water. The trophozoite, under adverse environmental stress, may encyst tolerating desiccation indefinitely. When the medium contains adequate nutritional components, excystment may take place. The trophozoite may emerge through the ostioles or pores.[51]

B. Life Cycle of *Acanthamoeba* spp. (Figure 16)

The trophozoite or vegetative form of *Acanthamoeba* sp. may form cysts when desiccation or other adverse conditions prevail, and can reverse to the vegetative or pro-

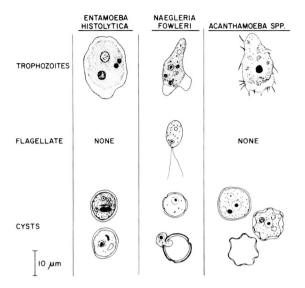

FIGURE 17.   Comparative morphology of amebas. (From Martinez, A. J. and Amado-Ledo, D. E., *Morphologa Normal y Patologica*, Sec. B, 3, 679, 1979. With permission.)

liferative form when the conditions are favorable.[58,86] Encystment may be induced by transferring trophozoites to a nonnutrient medium or adding metabolic inhibitors to the medium. Mature cysts can undergo excystment when returned to a growth medium.

C. Comparative Morphology of Amebas (Figure 17)

This diagram depicts and compares the basic similarities and differences between free-living or amphizoic and parasitic amebas.[112]

D. Temperature Tolerance

N. fowleri is a thermophilic free-living ameba growing well at temperatures up to 45°C. By contrast, *Acanthamoeba* sp. grow better at lower temperatures (25 to 35°C). Apparently, *Acanthamoeba* strains of low virulence are able to tolerate lower maximum temperatures than the pathogenic strains. Free-living amebas have been found at sea level, high altitudes (Titicaca Lake), in the Antarctica, and the fiords of Norway by T. J. Brown. In addition to the temperature, the age of the culture (days, weeks, or months, or number of passages) has a direct influence on the pathogenicity of the amebas; for example, the older the culture the longer the animals survive. The growth, multiplication, and maintenance of the different strains may also depend on the composition of the culture media.[34,38,113]

E. Mitosis and Chromosomes

N. fowleri, N. jadini, N. gruberi, and other *Naegleria* spp. divide by binary fission or promitosis with the presence of polar masses, polar caps, and interzonal bodies.[37] Nuclear division begins with a swelling of the nucleus and the elongation of the nucleolus. The nucleolus then divides into two equal halves in the prophase forming the polar masses, each of which comprises one half of the divided nucleolus. During metaphase the spindle apparatus may be seen connecting the polar masses. The chromatin then divides into two masses; each half moves toward its pole during anaphase. Then the polar caps disappear. The nuclear membrane persists during the nuclear division. At telophase the two daughter nuclei are formed. An interzonal body is usually seen during the separation in two of the nucleolus or chromosomal discs. Centrioles are not present in *Naegleria*.[51,59,121,122]

A. castellanii also divides by binary fission giving the appearance of fragmentation of the nucleolus. The nucleolus gradually disappears during prophase. During metaphase the chromatin assumes the shape of a solid band at the equator, which then divides in two. The centrosphere is an organelle present in the cytoplasm near the nucleus which may be seen at interphase and consists of a dense, elongated rod to which microtubules are attached. In anaphase the chromosomes reach the two ends of the poles and, finally, during telophase the nuclear membrane surrounds the two daughter nuclei. However, the chromosomes are not distinguishable, probably because they are very numerous, small, tenuous, and delicate. Centrioles are present in *Acanthamoeba* spp. during the division. Mitotic figures in the amebas are difficult to detect. Many transcriptional and translational events during encystation have been elucidated through the study of synthesis of actin and ribosomal proteins.[60]

Entamoeba histolytica has six chromosomes. *Amoeba proteus* has between 500 to 600. During the mitotic division of the *Naegleria* ameba, the chromosomes are barely stainable (modified Feulgen) at the light microscopic level and are not seen at the electron microscopic level. However, it is estimated that *N. gruberi* have 14 to 16 chromosomes.[100] It is estimated that the number of chromosomes in *Acanthamoeba castellanii* is approximately 80.[142] *Acanthamoeba* spp. either have a large number of very small chromosomes or a complex network of chromosomal material.[65a] For this reason, they are difficult to isolate and count. The mitotic pattern of *Tetramitus rostratus* appears to resemble that of *N. gruberi* since the nucleolus is dumbbell shaped and the typical chromosomes are not evident, but condensation of the nucleolar ground substance can be seen. *Hartmannella astronyxis* have approximately 14 to 22 short, rod-shaped chromosomes.[106]

V. IMMUNOLOGICAL AND ANTIGENIC CHARACTERISTICS OF FREE-LIVING AMEBAS

A. Immunoelectrophoretical Analysis and Its Variants

There are different proteins forming the cytoplasm of free-living amebas.[24,27] These proteins may be detected and characterized by different variants of immunoelectrophoresis.[52,61] The immunoelectrophoretic profiles are useful for taxonomy and classification of genera and species. For example, pathogenic *Naegleria* may be differentiated from nonpathogenic *Naegleria* by the isoenzyme pattern or zymogram method using the agarose isoelectric focusing.[39,134] In addition, other variations of the electrophoresis may be used such as immunodiffusion, polyacrylamide gel electrophoresis, immunoabsorption, starch gel electrophoresis; or zymograms of acid phosphatase, leucine amino peptidase, malate dehydrogenase, propionyl esterase, glucose phosphate, isomerase, and others.[39,43,44,89,157] All of these methods are reliable for species identification and taxonomy.[9,32,71,72]

Immunoelectrophoresis and its variants of the water-soluble protein extracts of free-living amebas permit identification and separation of the different species belonging to these genera. Immunological and antigenic differences between species have been found.[141] However, there is also significant cross-reactivity between them.[85] Indirect hemagglutination test, fluorescent antibody technique, and complement-fixation reaction have also been used to demonstrate amebic antigens and antibodies.[20,23,24] There are numerous recent reports and articles on immunological studies with detail of technical matters.[3-5,20,31,64,101-105,126,141,144-150] A major portion of the plasma membrane of *Acanthamoeba* spp. is made of lipophosphonoglycan which is an antigenic glycosphingolipid. Immunofluorescence may reveal antibodies present over the outer surface and the phagosomes of the trophozoites.[7]

A. castellanii, A. polyphaga, A. rhysodes, and *A. culbertsoni* are immunologically very closely related. Immunoelectrophoretic analysis demonstrates clear differences between *N. gruberi, N. fowleri, N. jadini,* and others.[7,152]

Indirect fluorescent-antibody testing on human sera, including samples from newborn children, was found fluorescent in all, either with neat serum or with serum diluted up to 1/20 of *Naegleria* and 1/80 of Acanthamoeba.[35] It has been postulated that normal human serum contains a natural antibody in low dilution which is capable of protecting the host in ordinary circumstances against infection. Cursons et al. postulated that since antibodies against free-living amebas are widespread in human serum, they might also represent cross-reactivity antibodies against a still unknown defined antigen.[35] It has been demonstrated that cell-mediated immunity may also play a part.[35] Probably different types of antibodies are involved in the antiamebic humoral response. IgA, IgG, and IgM are the immunoglobulins that may prevent the spread of the infection by helping to destroy the infecting trophozoites.

Cytopathogenicity of free-living amebas may be investigated by using tissue cell cultures or cultures with bacteria as food.[6,32,53,55-57,66,67,73,132,133] Differences in virulence between different genera have also been reported by observing cytopathic effects on cell cultures.[66,67] It has been demonstrated that cytolytic enzymes (acid hydrolases) may assist in the penetration of the host cells.[52]

Immunoglobulin levels may be elevated in animals infected with pathogenic *N. fowleri.*[136] This apparently induced protective immunity.

Complement-fixation tests from patients using antigen made from *A. culbertsoni* demonstrated increased antibody titers in some. These findings suggest the presence of natural antibodies in the normal population.[61] In addition to all of the methods used to study free-living and amphizoic amebas, the effects of hydrostatic pressure and concentration of gases (nitrogen and helium) toward the morphology of trophozoites have been investigated.[125] It has been demonstrated that absence of encystment, modification of morphology, and high mortality of the amebas depend on the concentration of such gases and their pressure in the water.[125]

B. Detection of Amebic Antigens and Antibodies in Body Fluids and Tissues — Potential Value for Diagnosis

The detection of amebic antigens, precipitins, and other hydrosoluble proteins and antibodies is a well-established procedure in the diagnosis of parasitic diseases and in *Entamoeba histolytica.* This can be achieved by several methods such as gas chromatography, latex agglutination, radio-immunoassay (RIA), double immunodiffusion (DID), counter immunoelectrophoresis (CIE), enzyme-linked immunosorbent assay (ELISA), immune adherence hemagglutination (IAHA), immunoperoxidase using specific antisera, etc. The use of these methods may be applied also for the diagnosis of free-living amebas. The RIA and ELISA offers the advantage of practicality, specificity, and sensitivity in identifying and quantifying amebic antigens.[5,62]

REFERENCES

1. Adam, K., The growth of *Acanthamoeba* sp. in a chemically defined medium, *J. Gen. Microbiol.,* 21, 519, 1959.
2. Adam, K., A comparative study of Hartmannellid amoebae, *J. Protozool.,* 11, 433, 1964.
3. Adam, K. and Blewett, D. A., Studies on the DNA of *Acanthamoeba, Ann. Soc. Belge Med. Trop.,* 54, 387, 1974.
4. Akins, R. A. and Byers, T. J., Differentiation promoting factors induced in *Acanthamoeba* by inhibitors of mitochondrial macromolecule synthesis, *Dev. Biol.,* 78, 126, 1980.

5. Alper, E. I., Litter, C., and Monroe, L. S., Counterelectrophoresis in the diagnosis of amebiasis, *Am. J. Gastroenterol.,* 65, 63, 1976.

6. Anderson, K. and Jamieson, A., Agglutination test for the investigation of the genus *Naegleria, Pathology,* 4, 273, 1972.

7. Anderson, K. and Jamieson, A., Bacterial suspensions for the growth of *Naegleria* species, *Pathology,* 6, 79, 1974.

8. Bailey, C. F. and Bowers, B., Localization of lipophosphonoglycan in membranes of *Acanthamoeba* by using specific antibodies, *Mol. Cell. Biol.,* 1, 358, 1981.

9. Bogler, S. A., Zarley, C. D., Burianek, L. L., Fuerst, P. A., and Byers, T. J., Interstrain mitochondrial DNA polymorphism detected in *Acanthamoeba* by restriction endonuclease analysis, *Mol. Biochem. Parasitol.,* 8, 145, 1983.

10. Bowers, B. and Korn, E. D., The fine structure of *Acanthamoeba castellanii.* I. The Trophozoite, *J. Cell Biol.,* 39, 95, 1968.

11. Bowers, B. and Korn, E. D., The fine structure of *Acanthamoeba castellanii* (Neff strain). II. Encystment, *J. Cell Biol.,* 47, 786, 1969.

12. Bowers, B., A morphological study of plasma and phagosome membranes during endocytosis in *Acanthamoeba, J. Cell Biol.,* 84, 246, 1980.

13. Byers, T. J., Growth, reproduction, and differentiation in Acanthamoeba, in *International Review of Cytology,* Vol. 16, Bourne, G. H. and Danielli, J. F., Eds., Academic Press, New York, 1979, 283.

14. Byers, T. J., Akins, R. A., Maynard, B. J., Kefken, R. A., and Martin, S. M., Rapid growth of *Acanthamoeba* in defined media; induction of encystment by glucose-acetate starvation, *J. Protozool.,* 27, 216, 1980.

15. Byers, T. J., Bogler, S. A., and Buriabrek, L. L., Analysis of mitochondrial DNA variation as an approach to systematic relationship in the genus *Acanthamoeba, J. Protozool.,* 30, 198, 1983.

16. Carosi, G., Etude comparative de l'ultrastructure d'*Entamoeba moshkovskii,* des amibes parasites du genre *Entamoeba* et des amibes "free-living" du groupe *Hartmannella-Naegleria, Ann. Soc. Belge Med. Trop.,* 54, 265, 1974.

17. Carosi, G., Scaglia, M., and Filice, G., An electron microscope study of *Naegleria jadini* Nov. sp. (Willaert-LeRay, 1973) in axenic "medium". The ameboid stage, *Protistologica,* 12, 31, 1976.

18. Carosi, G., Scaglia, M., Filice, G., Dei Cas, A., Carnevole, G., Gatti, S., and Benzi-Cipelli, R., Patterns of ultrastructural organization of free-living amoebae of *Acanthamoeba. Naegleria* group, *G. Malattie Infetive Parassitarie,* 29, 654, 1977.

19. Carosi, G., Filice, G., Scaglia, M., Gatti, S., and Torresani, P., Electron microscope study of *Acanthamoeba castellanii*-group spp. *A. castellanii, A. rhysodes, A. polyphaga, Rev. Parassitol.,* 39, 49, 1978.

20. Červa, L., Immunological studies on Hartmannellid amoebae, *Folia Parasitol. (Praha),* 14, 19, 1967.

21. Červa, L., Zimak, V., and Novak, M., Amoebic meningoencephalitis: a new amoeba isolate, *Science,* 163, 575, 1969.

22. Červa, L., Comparative morphology of three pathogenic strains of *Naegleria gruberi, Folia Parasitol. (Praha),* 17, 127, 1970.

23. Červa, L. and Kramar, J., Antigenic relationships among several limax amoebae isolates assessed with the indirect fluorescent antibody test (IFAT), *Folia Parasitol. (Praha),* 20, 113, 1973.

24. Červa, L., Detection of antibody against Limax amoebae by means of the indirect haemagglutination test, *Folia Parasitol. (Praha),* 24, 293, 1977.

25. Chambers, J. A. and Thompson, J. E., A scanning electron microscopic study of the excystment process of *Acanthamoeba castellanii, Exp. Cell Res.,* 73, 415, 1972.

26. Chang, S. L., Small free-living amoebas: cultivation, quantitation, identification, pathogenesis and resistance, *Curr. Top. Comp. Pathol.,* 1, 201, 1971.

27. Childs, G. E., *Hartmannella culbertsoni:* enzymatic ultrastructural and cytochemical characteristics of peroxisomes in a density gradient, *Exp. Parasitol.,* 34, 44, 1973.

28. Chiovetti, R., Re-encystment of the amoeboflagellate *Naegleria gruberi, Trans. Am. Micros. Soc.,* 95, 122, 1976.

29. Chiovetti, R., Encystment of *Naegleria gruberi* (Schardinger). I. Preparation of cysts for electron microscopy, *Trans. Am. Microsc. Soc.,* 97, 245, 1978.

30. Culbertson, C. G., The pathogenicity of soil amoebas, *Annu. Rev. Microbiol.,* 25, 231, 1971.

31. Culbertson, C. G. and Harper, K., Surface coagglutination with formalinized, stained protein a staphylococci in the immunologic study of three pathogenic amebae, *Am. J. Trop. Med. Hyg.,* 29, 785, 1980.

32. Cursons, R. T. M., Brown, T. J., and Keys, E. A., Immunity to pathogenic free-living amoebae, *Lancett,* 2, 875, 1977.

33. Cursons, R. T. M., Brown, T. J., and Keys, E. A., Virulence of pathogenic free-living amoebae, *J. Parasitol.,* 64, 744, 1978.

34. Cursons, R. T. M., Donald, J. J., Brown, T. J., and Keys, E. A., Cultivation of pathogenic and nonpathogenic free-living amebae, *J. Parasitol.*, 65, 189, 1979.
35. Cursons, R. T. M., Brown, T. J., Keys, E. A., Moriarity, K. M., and Till, D., Immunity to pathogenic free-living amoebae: role of humoral antibody, *Infect. Immun.*, 29, 401, 1980.
36. Daggett, P.-M. and Nerad, T. A., The biochemical identification of *Vahlkampfiid amoebae*, *J. Protozool.*, 30, 126, 1983.
37. Das, S. R., Willaert, E., and Jadin, J. B., Studies on mitotic division in *Naegleria jadini*, *Ann. Soc. Belge Med. Trop.*, 54, 141, 1974.
38. DeJonckheere, J., Use of an axenic medium for differentiation between pathogenic and nonpathogenic *Naegleria fowleri* isolates, *Appl. Environ. Microbiol.*, 33, 751, 1977.
39. De Jonckheere, J. and Van de Voorde, H., Comparative study of six strains of *Naegleria* with special reference to nonpathogenic variants of *Naegleri fowleri*, *J. Protozool.*, 24, 304, 1977.
40. De Jonckheere, J., Differences in virulence of *Naegleria fowleri*, *Pathol. Biol.*, 27, 453, 1979.
41. De Jonckheere, J., *Naegleria australiensis* sp. nov., another pathogenic *Naegleria* from water, *Protistologica*, 17, 423, 1981.
42. De Jonckheere, J., Isoenzyme patterns of pathogenic and non pathogenic *Naegleria* spp. using agarose isoelectric focusing, *Ann. Microbiol. (Inst. Pasteur)*, 133, 319, 1982.
43. De Jonckheere, J. and Dierickx, P. J., Determination of acid phosphatase and leucine amino peptidase activity as an identification method for pathogenic *Naegleria fowleri*, *Trans. R. Soc. Trop. Med. Hyg.*, 76, 773, 1982.
44. DeJonckheere, J., Isoenzyme and total protein analysis by agarose isoelectric focusing and taxonomy of the genus *Acanthamoeba*, *J. Protozool.*, 30, 701, 1983.
45. Dingle, A. and Fulton, C., Development of the flagellar apparatus of *Naegleria*, *J. Cell Biol.*, 31, 43, 1966.
46. Dingle, A. D., Control of flagellum number in *Naegleria*. Temperature shock induction of multiflagellate cells, *J. Cell Sci.*, 7, 463, 1970.
47. Dingle, A. D., Cellular and environmental variables determining numbers of flagella in temperature-shocked *Naegleria*, *J. Protozool.*, 26, 604, 1979.
48. Dunnebacke, T. and Schuster, F. L., Infectious agent from a free-living soil ameba *Naegleria gruberi*, *Science*, 174, 516, 1971.
49. Dunnebacke, T. and Schuster, F. L., The nature of a cytopathogenic material present in amebae of the genus *Naegleria*, *Am. J. Trop. Med. Hyg.*, 26, 412, 1977.
50. Fernández-Galiano, D., Silver impregnation of ciliated protozoa: procedure yielding good results with the pyridinated silver carbonate method, *Trans. Am. Microsc. Soc.*, 95, 557, 1976.
51. Fulton, C. and Guerrini, A. M., Mitotic synchrony in *Naegleria* ameba, *Exp. Cell Res.*, 56, 194, 1969.
52. Hadas, E., The lytic enzymes of pathogenic and nonpathogenic strains of *Acanthamoeba castellanii* and *Naegleria fowleri*, *Acta Protozool.*, 21, 111, 1982.
53. Haight, J. B. and John, D. T., Growth of *Naegleria fowleri* in several axenic media, *Folia Parasitol. (Praha)*, 27, 207, 1980.
54. Hawes, R. S. J., On *Rosculus ithacus* gen. n., sp.n. (protozoa, *Amoebina* with special reference to its mitosis and phylogenetic relations, *J. Morphol.*, 113, 139, 1963.
55. Howells, R. E., Saygi, G., and Warhurst, D. C., Observations on ageing cultures of amoebae, *Trans. R. Soc. Trop. Med. Hyg.*, 64, 19, 1970.
56. Hysmith, R. M. and Franson, R. C., Elevated levels of cellular and extracellular phospholipases from pathogenic *Naegleria fowleri*, *Biochim. Biophys. Acta*, 711, 26, 1982.
57. Hysmith, R. M. and Franson, R. C., Degradation of human myelin phospholipids by phospholipase-enriched culture media of pathogenic *Naegleria fowleri*, *Biochim. Biophys. Acta*, 712, 698, 1982.
58. Jadin, J. M., Eschbach, H. L., Verheyen, F., and Willaert, E., Etudes comparatives des kystes de *Naegleria* et d' *Acanthamoeba*, *Ann. Soc. Belge Med. Trop.*, 54, 259, 1974.
59. Jamieson, A. and Anderson, K., A simple method for studying nuclear division in free-living soil amoebae, *J. Clin. Pathol.*, 25, 271, 1972.
60. Jantzen, H., Schulze, I., Horstmann, U., and Christofori, G., Control of protein synthesis in *Acanthamoeba castellanii*, *J. Protozool.*, 30, 204, 1983.
61. John, D. T., Cole, T. B., and Marciano-Cabral, F. M., Sucker-like structures on the pathogenic amoeba *Naegleria fowleri*, *Appl. Environ. Microbiol.*, 47, 12, 1984.
62. Krupp, I. M., Comparison of counter immunoelectrophoresis with other serologic tests in the diagnosis of amebiasis, *Am. J. Trop. Med. and Hyg.*, 23, 27, 1974.
63. Lasman, M., Light and electron microscopic observations on encystment of *Acanthamoeba palestinensis* Reich, *J. Protozool.*, 24, 244, 1977.
64. Lastovica, A. J. and Dingle, A. D., Superprecipitation of an actomyosin-like complex isolated from *Naegleria gruberi* amoebae, *Exp. Cell Res.*, 66, 337, 1971.
65. Lastovica, A. J., Scanning electron microscopy of pathogenic and nonpathogenic *Naegleria* cysts, *Int. J. Parasitol.*, 4, 139, 1974.

65a. Linnemann, W. A. M., personal communication, May 28, 1980.

66. Marciano-Cabral, F. M., Patterson, M., John, D. T., and Bradley, S. G., Cytopathogenicity of *Naegleria fowleri* & *Naegleria gruberi* for established mammalian cell cultures, *J. Parasitol.*, 68, 1110, 1982.

67. Marciano-Cabral, F. and John, D. T., Cytopathogenicity of *Naegleria fowleri* for rat neuroblastoma cell cultures: scanning electron microscopy study, *Infect. Immun.*, 40, 1214, 1983.

68. Molet, B. and Kremer, M., Techniques d'etudes et criteres morphologiques por l'identification des amibes libres, *Bull. Soc. Sci. Vet. Med. Comparee*, 78, 215, 1976.

69. Molet, B. and Ermolieff-Braun, G., Description d'une amibe d'eaux douce: *Acanthamoeba lenticulata*, Sp. Nov *(Amoebida)*, *Protistologica*, 12, 571, 1976.

70. Neff, R. J., Purification, axenic cultivation and description of a soil amoeba, *Acanthamoeba sp.*, *J. Protozool.*, 4, 176, 1957.

71. Neff, R. J., Mechanisms of purifying amoeba by migration on agar surfaces, *J. Protozool.*, 5, 226, 1958.

72. Nerad, T. A. and Daggett, P.-M., Starch gel electrophoresis: an effective method for separation of pathogenic and nonpathogenic *Naegleria* strains, *J. Protozool.*, 26, 613, 1979.

73. Odell, W. D. and Stevens, A. R., Quantitative growth of *Naegleria* in axenic culture, *Appl. Microbiol.*, 25, 621, 1973.

74. Page, F. C., Taxonomic criteria for *Limax* amoebae with descriptions of 3 new species of *Hartmannella* and 3 of *Vahlkampfia*, *J. Protozool.*, 14, 499, 1967.

75. Page, F. C., Redefinition of the genus *Acanthamoeba* with descriptions of three species, *J. Protozool.*, 14, 709, 1967.

76. Page, F. C., *Hartmannella limax*, the original limax amoeba?, *Trans. Am. Microsc. Soc.*, 88, 199, 1969.

77. Page, F. C., Taxonomy and morphology of free-living amoebae causing meningoencephalitis in man and other animals, *J. Parasitol.*, 56, 257, 1970.

78. Page, F. C., Taxonomic and ecological distribution of potentially pathogenic free-living amoebae, *J. Parasitol.*, 56, 257, 1970.

79. Page, F. C., A further study of taxonomic criteria for *Limax* amoebae, with descriptions of new species and a key to genera, *Arch. Protistenkunde BD*, 116, S149, 1974.

80. Page, F. C., *Rosculus ithacus* Hawes, 1963 *(Amoebida, Flabelluidae)* and the amphizoic tendency in amoebae, *Acta Protozool.*, 12, 143, 1974.

81. Page, F. C., Morphological variation in the cyst wall of *Naegleria gruberi (Amoebida, Vahlkampfiidae)*, *Protistologica*, 11, 195, 1975.

82. Page, F. C., An illustrated key to freshwater and soil amoeba with notes on cultivation and ecology, *Freshwater Biol. Assoc. Sci. Publ.*, 34, 1, 1976.

83. Page, F. C., A revised classification of the *Gymnamoebia* (protozoa: Sarcodina), *Zool. J. Linnean Soc.*, 58, 61, 1976.

84. Page, F. C., A light- and electron-microscopical study of *Protacanthamoeba caledonica* n. sp., type-species of *Protacanthamoeba* n. g. *(Amoebida, Acanthamoebidae)*, *J. Protozool.*, 28, 70, 1981.

85. Pant, K. D., Prasard, B. N., and Singh, L. M., Antigenic relationship among fifteen strains of pathogenic *Hartmannella*, *Ind. J. Exp. Biol.*, 6, 227, 1968.

86. Pasternak, J. J., Thompson, J. E., Schultz, T. M. G., and Zachariah, K., A scanning electron microscopic study of the encystment of *Acanthamoeba castellanii*, *Exp. Cell Res.*, 60, 290, 1970.

87. Patterson, M., Woodworth, T. W., Marciano-Cabral, F., and Bradley, S. G., Ultrastructure of *Naegleria fowleri* enflagellation, *J. Bacteriol.*, 147, 217, 1981.

88. Pernin, P. and Pussard, M., Étude en microscopie photonique et électronique d'une amibe voisine du genre *Acanthamoeba: Comandonia operculata* n.gen., n.sp. *(Amoebida, Acanthamoebidae)*, *Protistologica*, 15, 87, 1979.

89. Pernin, P., Bouikhsain, I., De Jonckheere, J. F., and Petavy, A. F., Comparative protein patterns of three thermophilic nonpathogenic *Naegleria* isolates and two *Naegleria fowleri* strains by isoelectric focusing, *Int. J. Parasitol.*, 13, 113, 1983.

90. Proca-Ciobanu, M., Lupascu, G. H., Petrovici, A. L., and Ionescu, M. D., Electron microscopic study of pathogenic *Acanthamoeba castellanii* strain: the presence of bacterial endosymbionts, *Int. J. Parasitol.*, 5, 49, 1975.

91. Pussard, M., Cytologie d'une amibe terricole *Acanthamoeba terricola* n.sp., *Ann. Soc. Natl. Zool. (Paris)*, 6, 565, 1964.

92. Pussard, M., *Acanthamoeba comandoni* n.sp. comparaison avec *A. terricola* Pussard, *Rev. Ecol. Biol. Sol*, 1, 586, 1964.

93. Pussard, M., Le genre *Acanthamoeba volkonsky*, 1931 *(Hartmannellidae-Amoebida)*, *Protistologica*, 2, 71, 1966.

94. Pussard, M., Comparaison morphologique de 4 Souches d'*Acanthamoeba* du groupe *astronyxis-comandoni*, *J. Protozool.*, 19, 557, 1972.

95. Pussard, M., Modalités de la division nucléaire et taxonomie chez les amibes (*Amoebea*, protozoa). Révision des notions de promitose, mésomitose et métamitose, *Protistologica*, 9, 163, 1973.

96. Pussard, M., La morphologie des amibes libres. Intérêt et principes d'études, *Ann. Soc. Belge Med. Trop.*, 54, 249, 1974.

97. Pussard, M. and Pons, R., Morphologie de la paroi kystique et taxonomie du genre *Acanthamoeba* (protozoa, *Amoebida*), *Protistologica*, 13, 557, 1977.

98. Pussard, M., Tassonomie di amebe libere d'interesse medico, *G. Malatti Infettive Parassitarie*, 29, 668, 1977.

99. Pussard, M. and Pons, R., A study of the cystic pores of *Naegleria* (*Vahlkampfiidae-Amoebida*), *Protistologica*, 15, 163, 1979.

100. Rafalko, J. S., Cytological observations on the amoebo-flagellate, *Naegleria gruberi*, (Protozoa) *J. Morphol.*, 81, 1, 1947.

101. Raizada, M. and Krishna-Murti, C. R., Changes in the activity of certain enzymes of *Hartmannella* (Culbertson strain A-1) during encystment, *J. Protozool.*, 18, 115, 1971.

102. Raizada, M. K. and Mohan, R., L-Threonine dehydrase activity of axenically grown *H. culbertsoni*, *Arch. Microbiol.*, 85, 119, 1972.

103. Raizada, M. K. and Krishna Murti, C. R., Synthesis of RNA, protein, cellulose and mucopolysaccharide and changes in the chemical composition of *H. culbertsoni* during encystment under axenic conditions, *J. Protozool.*, 19, 691, 1972.

104. Raizada, M. K., Saxena, K. C., and Krishna Murti, C. R., Serological changes associated with the encystment of axenically grown *Hartmannella culbertsoni*, *J. Protozool.*, 19, 98, 1972.

105. Raizada, M. K. and Krishna-Murti, C. R., Binding of taurine to *H. culbertsoni* and the synthesis of cyclic AMP, *Curr. Sci.*, 42, 202, 1973.

106. Ray, D. L. and Hayes, R. E., *Hartmannella astronyxis:* a new species of free-living amoeba. Cytology and life cycle, *J. Morphol.*, 95, 159, 1954.

107. Rondanelli, E. G., Carosi, G., Filice, G., and DeCarneri, I., Studio microelettronico sulle modificazioni dell'organizzazione ultrastrutturale nelle fasi cistiche di *Hartmanella*, *G. Malattie Infettive Parassitarie*, 22, 633, 1970.

108. Rondanelli, E. G., Carosi, G., Gerna G., and De Carneri, I., Aspetti ultrastrutturale delle forme cistiche di *Hartmannella castellanii*, *Bull. Inst. Sieroterapico Milanese*, 49, 81, 1970.

109. Rondanelli, E. G., Carosi, G., Filice, G., and DeCarneri, I., Rilieve critici sull'organizzazione ultrastrutturale di un ceppo patogeno (A-1 Di Culbertson) di *Hartmannella castellanii* cultivato "in vitro", *Parassitologia*, 13, 339, 1971.

110. Rondanelli, E. G., Carosi, G., Scaglia, M., and DeCarneri, I., Ultrastructure of *Entamoeba moshkovskii* with particular regard to the presence of surface active lysosomes, *Int. J. Parasitol.*, 4, 331, 1974.

111. Rondanelli, E. G., Carosi, G., Scaglia, M., and Dei Cas, A., An electron microscope study of *Naegleria fowleri* (Carter, 1970) in anexic "medium". The ameboid stage, *Protistologica*, 12, 25, 1976.

112. Rondanelli, E. G., Carosi, G., Scaglia, M., Dei Cas, A., and Lotznicker, M. D., Ultrastructural pattern of the ameboid trophozoite of *Naegleria fowleri* and *Acanthamoeba astronyxis*, etiological agents of primary amoebic meningoencephalitis (PAME) in man. A comparative microelectronic study, *G. Malattie Infettive Parassitarie*, 28, 450, 1976.

113. Santillana, I., Madrigal Sesma, M. J., and Martínez-Fernández, A. R., Modificación experimental de la patogenia de una cepa de *Acanthamoeba* sp., *Rev. Ibérica Parasitol.*, Vol. Extra, p. 263, 1982.

114. Sawyer, T. K. and Griffin, J. L., A proposed new family, *Acanthamoebidae* N. Fam. (order amoebida), for certain cyst-forming filose amoebae, *Trans. Am. Microsc. Soc.*, 94, 93, 1975.

115. Saygi, G., An amoeba-agglutination test with *Acanthamoeba* (*Hartmannella*), *Trans. R. Soc. Trop. Med. Hyg.*, 63, 426, 1969.

116. Scaglia, M., Villa, M., Gatti, S., Strosselli, M., and Grazioli, V., *Entamoeba moshkovskii:* a new isolate from sewage sludges in Italy (correspondence), *Trans. R. Soc. Trop. Med. Hyg.*, 76, 703, 1982.

117. Schuster, F. L., Virus-like bodies in *Naegleria-gruberi*, *J. Protozool.*, 16, 724, 1969.

118. Schuster, F. L. and Dunnebacke, T. H., Formation of bodies associated with virus-like particles in the amoeba flagellate *Naegleria gruberi*, *J. Ultrastructural Res.*, 36, 659, 1971.

119. Schuster, F. L., Virus-like particles and an unassociated infectious agent in Amoebae of the genus *Naegleria*, *Ann. Soc. Belge Med. Trop.*, 54, 359, 1974.

120. Schuster, F. L., Ultrastructure of cysts of *Naegleria* sp. A comparative study, *J. Protozool.*, 22, 352, 1975.

121. Schuster, F. L., Ultrastructure of mitosis in the Amoeba-flagellate *Naegleria gruberi*, *Tissue Cell*, 7, 1, 1975.

122. Schuster, F. L., Small amebas and ameboflagellates, in *Biochemistry and Physiology of Protozoa*, 2nd ed., Lewandoswky, M. and Hutner, S. H., Eds., Academic Press, New York, 1979, 215.

123. Schuster, F. L. and Twomey, R., Calcium regulation of flagellation in *Naegleria gruberi*, *J. Cell. Sci.*, 63, 311, 1983.

124. Seilhamer, J. J. and Byers, T. J., ATPase-associated oligomycin resistance in *Acanthamoeba castellanii, J. Protozool.*, 29, 394, 1982.

125. Simitzis-LeFlohic, A. M., Barthelemy, L., and Chastel, C., Effets de la pression hydrostatique per se et des gas inertes sur une amibe libre du type limax: *Acanthamoeba culbertsoni, Pathol. Biol.*, 31, 11, 1983.

126. Singh, B. N., A culture method for growing small free-living amoeba for the study of their nuclear division, *Nature*, 165, 65, 1950.

127. Singh, B. N., Nuclear division in nine species of small free-living amoebae and its bearing on the classification of the order *Amoebida, Philos. Trans. R. Soc. London Ser. B*, 236, 405, 1952.

128. Singh, B. N. and Das, S. R., Studies on pathogenic and nonpathogenic small free-living amoebae and the bearing of nuclear division on the classification of the order *Amoebida, Philos. Trans. R. Soc. London Ser. B Biol. Sci.*, 259, 435, 1970.

129. Singh, B. N., Classification of amoebae belonging to the order *Amoebida* with special reference to pathogenic free-living forms, *Curr. Sci.*, 41, 395, 1972.

130. Singh, B. N. and Hanumaiah, V., Studies on pathogenic and nonpathogenic amoeba and the bearing of nuclear division and locomotive form and behaviour on the classification of the order *Amoebida*, in 20th Annual Conf. of the AMI, Monograph #1, Association of Microbiologists of India. Haryana Agricultural University, Hissar, 1979.

131. Singh, B. N., Nuclear division as the basis for possible phylogenic classification of the order *Amoebida kent*, 1880, *Ind. J. Parasitol.*, 5, 133, 1981.

131a. Singh, B. N., personal communication, October 14, 1983.

132. Stevens, A. R. and O'Dell, W. D., The influence of growth medium on axenic cultivation of virulent and avirulent *Acanthamoeba, Proc. Soc. Exp. Biol. (N.Y.)*, 143, 474, 1973.

133. Stevens, A. R. and O'Dell, W. D., In vitro growth and virulence of *Acanthamoeba, J. Parasitol.*, 60, 884, 1974.

134. Stevens, A. R., Biochemical studies of pathogenic free-living amebae, *G. Malatti Infettive Parassitarie*, 29, 690, 1977.

135. Stratford, M. P. and Griffith, A. J., Variations in the properties and morphology of cysts of *Acanthamoeba castellanii, J. Gen. Microbiol.*, 108, 33, 1978.

136. Thong, Y. H., Carter, R. F., Ferrante, A., and Rowan-Kelly, B., Site of expression of immunity to *Naegleria fowleri* in immunized mice, *Parasite Immunol.*, 5, 67, 1983.

137. Tomlinson, G. and Jones, E. A., Isolation of cellulose from the cyst wall of a soil amoeba, *Biochim. Biophys. Acta*, 63, 194, 1962.

138. Van Saanen, M., Comparative study with the scanning electron microscope of cysts of *Naegleria* and *Acanthamoeba, Experientia*, 33, 1680, 1977.

139. Vickerman, K., Patterns of cellular organization in *Limax amoebae*. An electron microscope study, *Exp. Cell Res.*, 26, 497, 1962.

140. Visvesvara, G. S. and Balamuth, W., Comparative studies on related free-living and pathogenic amoebae with special reference to *Acanthamoeba, J. Protozool.*, 22, 245, 1975.

141. Visvesvara, G. S. and Healy, G. R., Comparative antigenic analysis of pathogenic and free-living *Naegleria* species by the gel diffusion and immunoelectrophoresis techniques, *Infect. Immun.*, 11, 95, 1975.

142. Volkonsky, M., *Hartmannella castellanii* Douglas et classification des Hartmannelles, *Arch. Zool. Exp. Gen.*, 72, 317, 1931.

143. Walsh, C., Mar, J., and Ugen, K., Induction of multiple flagella in *Naegleria*: requirements for FRN and protein synthesis, *Dev. Genet.*, 1, 2, 133, 1979.

144. Warhurst, D. C. and Thomas, S. C., An isoenzyme difference between a "smooth" (s) and "rough" (r) strain of *Naegleria gruberi, Protistologica*, 14, 87, 1978.

145. Weik, R. R. and John, D. T., Quantitation and cell size of *Naegleria fowleri* by electronic particle counting, *J. Parasitol.*, 63, 150, 1977.

146. Weik, R. R. and John, D. T., Cell size, macromolecular composition, and O_2 consumption during agitated cultivation of *Naegleria fowleri, J. Protozool.*, 24, 196, 1977.

147. Weik, R. R. and John, D. T., Agitated mass cultivation of *Naegleria fowleri, J. Parasitol.*, 63, 868, 1977.

148. Weik, R. R. and John, D. T., Macromolecular composition and nuclear number during growth of *Naegleria fowleri, J. Parasitol.*, 64, 746, 1978.

149. Weik, R. R. and John, D. T., Cell and mitochondria respiration of *Naegleria fowleri, J. Parasitol.*, 66, 700, 1979.

150. Weik, R. R. and John, D. T., Preparation and properties of mitochondria from *Naegleria gruberi, J. Protozool.*, 26, 311, 1979.

151. Weisman, R. A., Differentiation in *Acanthamoeba castellanii, Annu. Rev. Microbiol.*, 30, 189, 1976.

152. Willaert, E., Isolement et culture in vitro des amibes du genre *Naegleria, Ann. Soc. Belge Med. Trop.*, 51, 701, 1971.

153. Willaert, E., Jadin, J. B., and LeRay, D., Structures immunochimiques comparées d'amibes du genre *Naegleria, Protistologica*, 8, 497, 1972.
154. Willaert, E., Jadin, J. B., and LeRay, D., Comparative antigenic analysis of *Naegleria* species, *Ann. Soc. Belge Med. Trop.*, 53, 59, 1973.
155. Willaert, E. and LeRay, D., Caractères morphologiques, biologiques et immunochemiques de *Naegleria jadini* sp.nov. *(Amoebida, Vahlkampfiidae), Protistologica*, 9, 417, 1973.
156. Willaert, E., Immunotaxonomy of the genera *Naegleria* and *Acanthamoeba* and its diagnostic consequences in cases of amoebic meningoencephalitis, *G. Malattie Infettive Parassitarie*, 29, 680, 1977.
157. Woodworth, T. W., Keefe, W. E., and Bradley, S. G., Characterization of the proteins of *Naegleria fowleri*: relationship between subunit size and charge, *J. Protozool.*, 29, 246, 1982.

Chapter 4

ECOLOGY, EPIDEMIOLOGY, AND ENVIRONMENTAL FACTORS

I. ECOLOGICAL ASPECTS: GEOGRAPHICAL DISTRIBUTION

Strains of soil and water amphizoic amebas are widespread in nature. It is, therefore, appropriate to study the interrelations between these protozoa in the environment where they live and multiply.

They were thought to be nonpathogenic to animals and man until Culbertson demonstrated that strains of *Acanthamoeba* were capable of producing acute meningoencephalitis in experimental animals.[35,36] At first, reports attributed these infections to species of *Acanthamoeba* or *Hartmannella;* however, later investigations showed that species of *Naegleria* were involved in human cases.[3-6,9,12,13,38-41] About 130 cases have been reported world-wide due to *N. fowleri*[1,18,21-23,37,64,92,100,103,113,121,135] and more than 30 due to *Acanthamoeba* spp.[2,7,10,11,105] (Figure 1 and Table 1). Of these, 50 PAM and 19 GAE cases are from the U.S., mainly in the coastal areas (Figure 2 and Table 2).

Pathogenic free-living amebas appear more frequently in thermally enriched water collections, thermally polluted discharge water from industrial plants and lakes or swimming pools, while nonthermally enriched water reserves contain fewer amebas. Water at an elevated temperature may serve as a potential reservoir for perpetuation and possible spread of pathogenic and thermophilic free-living amebas. When the temperature exceeds 30°C pathogenic *Naegleria* sp. proliferate better.[17,27-31,39,43,45,49,51,56-58,60,64,68,69,84-86,96,109,140,148,150,155,161-164]

In the 16 cases reported from Czechoslovakia which occurred between 1962 to 1965, 10 were from different localities, but all used the same indoor swimming pool filled with river water that had been heated and chlorinated.[21-23] Virulent *N. fowleri* was found again in the same indoor swimming pool in 1977 and 1978.[89,90] Forms of low virulence and nonvirulence to experimental animals were isolated from a crack in the wall of the pool.[91]

In South Australia 14 cases of PAM have been reported, 3 in western Australia, 1 in Queensland, and 1 in New South Wales. The first eight cases occurred between 1961 and 1972 in Port Augusta. In Kadina, one case occurred in 1965 and another in 1969. In Port Pirie, three cases occurred between 1955 and 1971. From February, 1972 and January, 1981 there were no cases of PAM. The last case from South Australia occurred in January 1981.[63] An epidemiological survey disclosed that water heated by the sun in overland pipelines was the source of pathogenic free-living amebas in some of these cases. Intranasal inoculation while swimming, sprinkling water into the nose, playing submarine, or washing the nose and aspirating water through the nostrils were the most likely mechanisms of infection. These cases occurred during the summer months when the temperature reached high levels. The cases were confined around northern St. Vincent and Spencer Gulf regions (see Figure 7).

The majority of PAM and GAE cases have been reported from the U.S. (Table 2). The cases of PAM in Virginia came from a small area south of Richmond in Chesterfield County within a 10-km radius. The first case of PAM reported from Virginia occurred in July 1937 and the second in 1950. Three cases occurred in 1951 and four in 1952. No new cases were reported until 1957 when two cases were found. Then 11 years later PAM was again detected in 1966. In 1967 two cases were reported; in 1968 one case was found, and then in 1969 two more cases were discovered. The cases in Virginia occurred sporadically and in clusters, all of them during the hottest summer months.

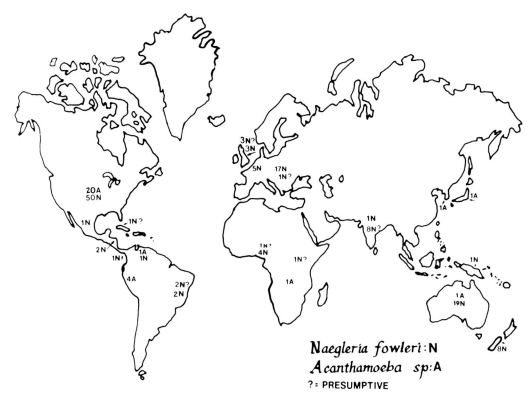

FIGURE 1.   Primary amebic meningoencephalitis (PAM) and granulomatous amebic meningoencephalitis (GAE). World-wide up to December 31, 1983.

Table 1
PAM AND GAE
(WORLD-WIDE — DECEMBER 31, 1983)

Country	GAE # cases	PAM # cases	Total #	Country	GAE # cases	PAM # cases	Total #
Proven cases[a]				Presumptive cases[b]			
Australia	1	19	20	Brazil	—	2	2
Belgium	—	5	5	Cuba	—	1	1
Brazil	—	2	2	Great Britain	—	3	3
Czechoslovakia	—	17	17	Hungary	—	1	1
Great Britain	—	3	3	India	—	8	8
Honduras	2	—	2	Uganda	—	1	1
India	1	—	1				
Japan	1	—	—	Total	33	129	162
Mexico	—	1	1				
New Guinea	—	1	1				
New Zealand	—	8	8				
Nigeria	1	4	5				
Panama	—	1	1				
Peru	4	—	4				
South Korea	1	—	1				
U.S.	20	50	70				
Venezuela	1	2	3				
Zambia	1	—	1				

[a]   Confirmed by autopsy/biopsy and/or culture with isolation of amebas.
[b]   Not confirmed by autopsy and/or culture of amebas.

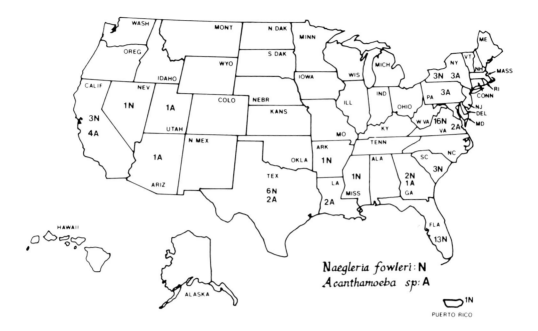

FIGURE 2. Primary amebic meningoencephalitis (PAM) and granulomatous amebic meningoencephalitis (GAE). In the U.S. up to December 31, 1983.

Table 2

PRIMARY AMEBIC MENINGOENCEPHALITIS (PAM) AND GRANULOMATOUS AMEBIC ENCEPHALITIS (GAE) (U.S. — DECMEBER 31, 1983)

State	GAE # of cases	PAM # of cases	Total #
Virginia	2	16	18
Florida	—	13	13
Texas	2	6	8
California	4	3[a]	7
New York	3[b]	3[c]	6
South Carolina	1	3	4
Georgia	1	2	3
Arkansas	—	1	1
Mississippi	—	1	1
Nevada[d]	—	1	1
Puerto Rico	—	1	1
Arizona	1	—	1
Louisiana	2	—	2
Pennsylvania	3	—	3
Utah[e]	1	—	1
Total	20	50	70

[a] One of these patients swam in lakes in Colorado and Oklahoma. MMWR 1980, 29, 405.

[b] One of these cases is suggested by serological data. Kenney, M., *Health Science Laboratory*, 1971, 8, 5.

[c] One of these patients swam in an East Central Florida lake. MMWR 1980, 29, 405.

[d] Swam in a desert lake near Las Vegas. Rothrock, J. F. and Buchsbaum, H. W., JAMA 1980, 243, 2329.

[e] Patient lived in Idaho. Grunnet, M. L., Cannon, G. H., and Kushner, J. P., *Neurol.*, (NY) 1981, 31, 174.

FIGURE 3.   Aerial view of Lake Chester located about 10 mi south of Richmond and close to Highway 95 that runs North-South. (From Martinez, A. J., et al., *Pathol. Annu.*, 12(2), 225, 1977. With permission.)

In that area there were several small ponds and man-made lakes. These lakes were located on both sides of the North-South Highway 95: Lake Chester was built in 1927 to 1928 when water from a well was poured into it in 1929 to 1930 (Figure 3). Lake Moore was built in 1929 and redone after 1951 (Figure 4). Public health authorities closed Lake Chester (Figures 5 and 6). Pictures taken in September 1970 demonstrated how shallow, muddy, and sandy the bottom of that lake actually was. Lake Moore is still operational (January 1984). Some of these lakes had inadequate water circulation or replenishment. The filtration of water was nonexistent and there were abundant amounts of decaying organic matter, mud, sand, and soil. Of course, these lakes were not chlorinated, or if they were, the chlorination was insufficient. During the weekend and holidays, they were overcrowded with swimmers which added organic matter and pollution to the water (see Table 1 in Chapter 7).

Eight cases due to *N. fowleri* have been reported from New Zealand.[17,37-41,43] All of these cases occurred in the North Island region near the 40th south parallel (Figure 8). All the fatalities in New Zealand occurred after the victims swam in natural thermal water either in a commercially developed basin, a natural pool, or river. Since the temperature of the thermal waters vary slightly, there was a lack of seasonal pattern of ameba isolation in New Zealand. Therefore, the eight cases were randomly distributed throughout the year. It had been suggested that heavy rainfall could increase the incidence of amebas in pools by increasing runoff from the surrounding soils. Some PAM cases of New Zealand PAM occurred after heavy rains.[17]

Five cases reported from Belgium also had a history of swimming in a heated swimming pool.[55,56] In Nigeria, cases of PAM unrelated to swimming caused by dust-borne amebas have been reported.[1,102,103]

In the city of Bath, Southwest England, the hot springs there were found contaminated by pathogenic *N. fowleri*. A major engineering project was carried out after an 11-year-old girl died of PAM while swimming there. The swimming areas in the Beau Street Bath, Hot Bath, Cross Bath, and the Kings Bath were rebuilt.[8]

FIGURE 4. Aerial view of Lake Moore. Vegetation is profuse. Highway 95 is also near this man-made lake. (From Martinez, A. J., et al., *Pathol. Annu.*, 12 (2), 225, 1977. With permission.)

FIGURE 5. Lake Chester after closure by the Public Health Department (September 1970).

In addition, fatal cases of PAM and GAE, confirmed by autopsies, have been reported in the following countries: South Korea, Japan, Peru, Venezuela, India, Zambia, Brazil, Mexico, Puerto Rico, and Panama. Cases of both diseases have also been reported, but not confirmed by autopsy, in Brazil, Hungary, Uganda, Cuba, and India (Table 1).

FIGURE 6.   Lake Chester after closing. The bottom is shallow and made up of sand and mud.,

FIGURE 7.   Map of Australia showing the areas of southern and western Australia where PAM cases occurred. Insert: Closer view of South Australia near the St. Vincent Gulf demonstrating the overland pipeline that carries the water to several towns (broken lines from Morgan to Whyalla; River Murray is on the right). (From Dorsch, M. M., Cameron, A. S., and Robinson, B. S., *Trans. Royal Soc. Trop. Med. Hyg.,* 77(3), 372, 1983. With permission.)

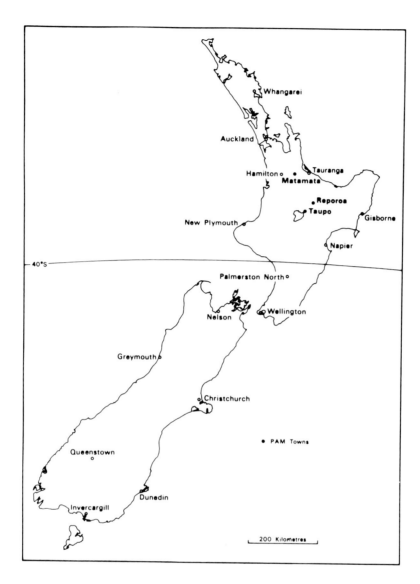

FIGURE 8. Map of New Zealand indicating the towns (black circles) where cases of PAM have been reported. Six cases occurred at Matamata; one at Reporoa and one at Taupo. (Courtesy of Dr. Tim J. Brown.)

Pathogenic amebas appear to proliferate and multiply better during the hotter (summer) months of the year suggesting a seasonal pattern. These strains are rare, transient, or ephemeral.[33,135,136,138] The summer months of the year are the high-risk period because water-sport activities are more frequent and the warm water stimulates the growth of pathogenic ameba.[67]

Isolation of pathogenic and nonpathogenic free-living amebas *(Naegleria* spp. and *Acanthamoeba* spp.) has been achieved world-wide (Antarctica; frozen swimming areas of Norway;[14,15,61] swimming pools in Stockholm and Berlin; rivers, lakes, and soil in Costa Rica and the Philippines; thermal mud samples from a spa in Italy; mountain water in Spain, France, Poland, Roumania, Canada, Brazil; sea level, ocean sediment, and high altitudes such as Lake Titicaca; New Zealand; Australia and India).[26,29,34,38,39,45,46,51,55-60,62,69-71,75-78,81,93-96,101,104,106-112,114,117,122-126,131-135,139,147,148,150,152-155,161-164]

Brown and his team from Massey University on an expedition to Antarctica were able to collect soil and water samples from the McMurdo Sound-Dry Valley region. They isolated and cultured free-living amebas from 22 of 70 samples they collected. These amebas grew well at 30°C but were not pathogenic for mice. The isolates of *Acanthamoeba* spp. have a better survival potential then the isolates of *Naegleria* spp.[15] Cursons and Brown recommended an excellent flow diagram for the general procedure of isolation and identification of free-living amebas.[38] They also isolated *N. fowleri* and *Acanthamoeba* spp. from several water samples from thermal pools and streams in the North Island of New Zealand.[17] These water samples also contained a high concentration of coliform bacteria along with soil contamination.

Scaglia and colleagues isolated and cultured free-living amebas from 160 samples from waters in Pavia, Italy. They identified free-living amebas in 75 samples (46.9%). Only six *Acanthamoeba* and two *Naegleria* were pathogenic for mice and grew well at 37°C (*Acanthamoeba* spp.) and 45°C (*Naegleria* spp.)[132,133] Martinez collected water from Lake Titicaca in Bolivia, (12,507 ft or 3,812 meters above sea level) and nonpathogenic *Acanthamoeba* spp. were isolated by J. Sykora (personal communication).

All pathogenic *N. fowleri* can grow well at high temperatures (40° to 45°C), but not all high-temperature isolates of free-living amebas are necessarily pathogenic.

In Florida, pathogenic *N. fowleri* were isolated from freshwater lakes showing that thermal discharge pollution of water plays a minor role in the multiplication of free-living amebas in semitropical areas.[161] Pathogenic *Acanthamoeba* spp. have also been isolated from ocean sediment.[131]

Nonpathogenic *Naegleria* spp. were isolated from soil and litter in Michigan. During the hottest summer months of August and September, trophozoites were more abundant than cysts, suggesting that during the coldest months *Naegleria* spp. remain predominantly in the cystic form.[152]

II. INCIDENCE AND PREVALENCE OF THE DISEASE: MORBIDITY AND MORTALITY

Free-living amebas are distributed world-wide, but the true incidence and prevalence is unknown. The disease has existed for a long time, because cases have been diagnosed retrospectively.[151] Because *Naegleria fowleri* is a thermophilic ameba, cases of PAM are more likely to occur during the hottest summer months. It is estimated that the risk of contracting PAM is in the range of 1 in 2.5 million "exposures" to contaminated water.[161] *Acanthamoeba* spp. is an opportunistic infection and can occur at anytime of the year without exposure to water. It has been suggested that the pathogenic species of free-living amebas are mutant, arising from *N. gruberi* stock. Perhaps living in polluted water and competing with other protozoa which develop well at higher temperatures, these free-living amebas undergo genetic changes influencing their virulence and pathogenicity.

Free-living amebas have been found in the nasal mucosa and oropharynx of "healthy" individuals. This may be due to transportation of the trophozoites or cysts by the air.[25,32,97,102,157-160] Free-living amebas have also been isolated from the surface of edible mushrooms.[118] There are several reports of contamination of tissue and bacterial culture by free-living amebas.[19,31,72,74,116] Dialysis units and air-conditioning units have been reported contaminated by *Acanthamoeba*[20,66,149] The possibility that free-living amebas are "carriers" of pathogenic bacteria has also been raised.[66,74,80-82,99,128,129]

PAM affects males about three times more often than females. This may be indicative of greater participation by males in water-related sports rather than sexual or racial predisposition. GAE, because of its opportunistic characteristics, may occur at anytime of the year and, of course, without exposure to fresh water.

Therefore, there is no specific susceptibility to *N. fowleri* infection, but there is a definite susceptibility to *Acanthamoeba* spp. in individuals with compromised defense mechanisms or chronic debilitating illness.[2,7,10] The mortality rate of both PAM and GAE is nearly 100%. There are no reported survivors of GAE. There are at least three patients who survived PAM with no sequelae because of rapid treatment.

A. Primary Amebic Meningoencephalitis (PAM) due to *Naegleria fowleri*

Swimming pools and man-made lakes are the principal habitat of *N. fowleri*. Elevated temperatures during the summer months or near the discharge outlets of power plants have been shown to facilitate the growth of pathogenic *Naegleria* spp. Victims are healthy, young individuals with a recent history of water-related sport activities.[153] The majority of PAM cases have occurred during the hottest season of the year because *N. fowleri* are thermophilic and proliferate better at higher temperatures. The chemical composition of the water, temperature, pH, and amount of organic matter all affect the multiplication of free-living amebas.[11,120] Iron-chelating agents of microbial origin have been reported to exert an inhibitory effect on *N. fowleri*.[87,88,98] As pathogenic *N. fowleri* are thermophilic, they proliferate or multiply better at higher temperatures. Temperatures as high as 45°C and as low as 26.5°C have been reported from which *N. fowleri* were isolated.[27,39,43,140,150] On the other hand, *A. castellanii* have been found in water from frozen lakes at 2°C. *A. culbertsoni* were also found at higher temperatures such as 42°C. A more accurate taxonomical classification may be made by use of specific antibodies.[73]

B. Granulomatous Amebic Encephalitis (GAE) due to *Acanthamoeba* spp.

Acanthamoeba sp. is also a widespread protozoon which can be found as "normal" flora in healthy individuals and can be isolated from samples of fresh water, air, bottled mineral water,[127,130] and even from dialysis machines or air-conditioning systems, but the infection produced is probably "opportunistic" because it attacks preferentially the chronically ill, and immunologically compromised individuals.[156]

III. PROPHYLAXIS, CONTROL, AND PREVENTION: INDIVIDUAL AND COMMUNITY MEASURES

Free-living amebas are ubiquitous protozoa and are widely dispersed in our environment. Avoiding exposure to contaminated water is an excellent preventive measure, as is treating recreational water with a sufficient amount of chlorine. On the basis of epidemiologic and clinical data and autopsy findings, well-documented risk factors have been identified. Subclinical infections are probably common in healthy people, with the protozoa having a precarious hold in the nose and throat.[65,99,102,115,136,137] It may be assumed that air contamination with cysts was the probable route of contamination. Free-living amebas have been isolated from the nose and throat of healthy soldiers, but the isolation of amebas was higher in individuals with rhinitis, headaches, and nose bleeds.[141-146]

The South Australia Health Commission has conducted campaigns to educate the public regarding quality and purification of drinking and recreational waters and proper maintenance of swimming and wading pools through television, radio, and pamphlets. Also, important legislative regulations have been implemented for disinfection and limiting access to swimming pools. The Ameba Monitoring Program was established to locate possible sources of contamination of supplies and to monitor the effectiveness of chlorination.[16] Water was sampled routinely to determine the free-chlorine residual, the total chlorine residual, the total coliform and *Escherichia coli* count, and to identify the ameba if present. Stimulation of research to find an effective treatment for PAM has been emphasized (Figures 9 to 13).

HOME SWIMMING POOLS

CARE, MAINTENANCE, DISINFECTION AND USE

Pool water for safe swimming needs regular attention to keep it clean, filtered and disinfected.
Keep pool surrounds and surface clean. Vacuum sediments from the pool daily and remove debris by skimming.

Pool Water Quality

To maintain pool water quality follow the instructions of the pool manufacturer, pool equipment and chemical suppliers.

- Maintain pH of the water (7.0-7.6)
- Flocculate as needed.
- Filter the water before and during use.
- Maintain the filter by back flushing or cartridge replacement.
- Disinfect the pool water as needed, frequency of dosing is related to usage and weather conditions.
- Prevent algal growth by proper disinfection and/or use of an algicide.
- Maintain the clarity of the water so the pool bottom is readily visible.
- Pool water chemicals can be hazardous if not handled with care. Follow label instructions.

Pool Water Testing

- Test level of disinfectant often as needed to maintain correct level.
- Use a reliable test kit and fresh test chemicals and follow instructions.

Personal Hygiene

- Do not spit, blow nose or urinate in water.
- Person's suffering with any infection should not use the pool.
- The use of suntan and body lotions create an oily scum and make pool water treatment difficult.

Further Information

Advice about care and maintenance of swimming pools is available from the South Australian Health Commission, Health Surveying Services, 158 Rundle Mall, (218 3629) or your Local Health Surveyor at your Council Office.

SWIM IN CLEAN WATER!

Issued by **South Australian Health Commission—October, 1981.**

FIGURE 9.   What you need to know about PAM. (Courtesy of the South Australian Health Commission.)

Free-living amebas have been isolated from bottled mineral water, home dialysis units, disposable filters attached to heating and ventilation, air-conditioning units, home-heated humidifiers, cooling ponds and towers of electric power plants, cooling systems of thermal and chemical power stations, nuclear power stations and aquariums

How can Health Professionals help minimize the incidence of Amoebic Meningitis?

1. Encourage everyone in the community, especially parents, to co-operate in seeing that these safety precautions are carefully observed:

- Keep your head above the water when swimming and if you must jump in—be sure to hold your nose.

- Ensure that swimming pools are correctly disinfected—local health surveyors will provide advice on proper pool maintenance.

- Empty and clean small collapsible pools daily.

- Keep your head above the water in the bath.

- Run the tap for a few minutes before use if you have been away.

- Keep sprinklers and hoses away from noses.

2. Display and distribute promotional material.

Left: The symbol of the Health Commission's publicity campaign to 'Prevent Amoebic Meningitis'

Reference:

Thong, Y. H. Primary Amoebic Meningoencephalitis: Fifteen Years Later. Med. J. Aust. 1980, 1:352-354
D. J. WOOLMAN GOVERNMENT PRINTER SOUTH AUSTRALIA

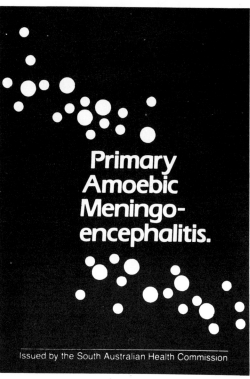

What you need to know about

Primary Amoebic Meningo-encephalitis.

Issued by the South Australian Health Commission

FIGURE 10.   Home swimming pools. (Courtesy of the South Australian Health Commission.)

from homes, laboratories, or zoological gardens.[90] The control of organic matter and pollution of water in swimming pools is an excellent preventive measure. Routine treatment of the water by chlorine and other disinfectants could prevent the multiplication of free-living amebas to some extent. Chlorine is more effective against *Naegleria* spp. than *Acanthamoeba* spp.[16] Iodine application in concentrations from 0.076 to 0.50 mg/ℓ showed no lethal effect on the amebas.[24] The addition of chemicals to the water (e.g., sodium chloride, aluminum sulfate, iodine disinfectants) or other competitive or predator protozoa or bacteria may be detrimental or beneficial to free-living amebas. Therefore, more research is needed. Apparently, there is no human carrier of pathogenic *N. fowleri* — neither asymptomatic nor subclinical infections. Natural antibodies may be present in some individuals protecting them.[42]

Baquacil has been tested and proven to be effective as a disinfectant.[47,48] Physical agents such as drying or dessication heat and low temperature have also been tested.

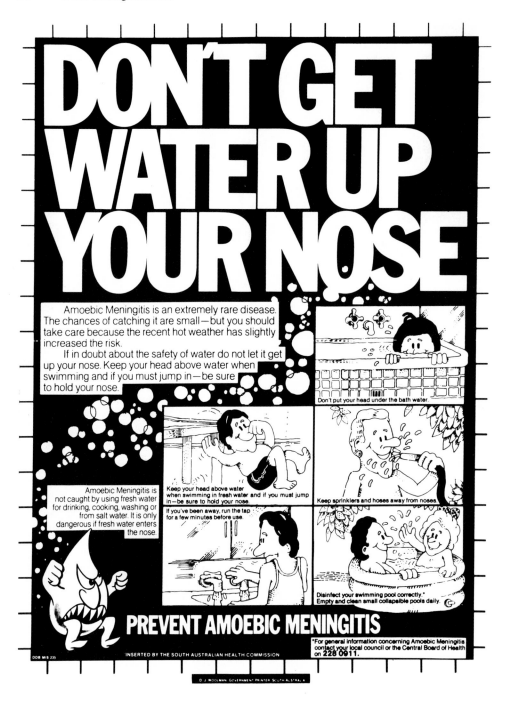

FIGURE 11.   Don't get water up your nose. (Courtesy of the South Australian Health Commission.)

Drying and freezing make trophozoites become cysts that may survive frozen. In addition, Deciquam 222, ozone, chlorine, and chlorine dioxide have all been shown to possess lethal effects against free-living amebas in higher concentrations than used to kill bacteria.[44]

However, it is puzzling that PAM due to *N. fowleri* is not more frequently found, particularly in areas with a tropical climate. Generally, all patients with PAM had a

FIGURE 12. Follow these rules to prevent PAM. (Courtesy of South Australian Health Commission.)

history of swimming in fresh-water lakes or ponds a few days before onset of the symptoms. It is known that these amebas grow better at higher temperatures.[27] Apparently, chlorination of the water does not entirely eliminate the presence of pathogenic strains.[44,49] In fact, cases of PAM occurred where swimming pools were treated with chlorine. Apparently, the amount was inadequate to eliminate both normal flora and the pathogenic amebas. This may lead to a disruption of the biological balance between the nonpathogenic (less resistent) and the pathogenic (resistant).[51,119,120]

Experimental work has demonstrated that 0.5 mg/ℓ of free-chlorine residual at pH of 7 is sufficient to kill about 99% of the amebic trophozoites provided the contact time is at least 30 min.[44]

Significant data have been collected on the effects of chemicals on amebic trophozoites and cysts. Total elimination of free-living amebas from swimming pools is a

RUN TAPS BEFORE USE

All water supplies have been fully chlorinated — but the level may drop in the house pipes if the water has not been used for a long time. RUN THE TAP UNTIL YOU CAN SMELL THE CHLORINE — PARTICULARLY IF YOU HAVE BEEN AWAY.

SWIM IN SALT WATER ONLY

Swimming in the SEA or in SALT-WATER SWIMMING POOLS (70 lbs. per 1,000 gallons min.) is safe. All swimming pools should be continuously filtered and chlorinated as usual for other health reasons. DON'T SWIM IN FRESH WATER, DAMS or TANKS.

DON'T JUMP IN

Jumping in is the surest way to force water up the nose. So in fresh water DO NOT JUMP IN. Jumping and diving in looks daring and attractive — but it can be terribly dangerous.

PADDLING POOLS CAN BE DANGEROUS

Small collapsible plastic and canvas paddling pools are difficult to keep clean — and cysts in the soil and grass may be carried into the water on the feet. Paddling pools should be emptied daily and refilled with salty water (1½ ozs. salt per gallon minimum). The water should also be chlorinated where possible.

DON'T PLAY "SUBMARINES" IN THE BATH

Heads down and tails up is all right for ducks. For humans it is more than likely that water will be forced up the nose. Keep your tail down and your head up in the bath.

DON'T PLAY WITH SPRINKLERS AND HOSES

On a hot day the garden sprinkler is beautifully cool and refreshing. It could kill you — if the water has active amoebae in it and if the water is forced into your nose. By all means play — but play safe.

IN OTHER WORDS — KEEP YOUR HEAD ABOVE WATER

FIGURE 13.   Keep your head above water. (Courtesy of the South Australian Health Commission.)

difficult and almost impossible task, particularly since amebas and cysts are introduced into the water from the air or through human carriers.[50,77,78,92,93]

Swimming in lakes, rivers, or swimming pools polluted by sewage and organic matter should be avoided. Anyone entering swimming pools should take a shower and wash their feet before entering the swimming pool. They should also take a shower after swimming to eliminate epithelial cells from their bodies. Hyperchlorination of the water has been suggested, but such treatment would be as unacceptable to swimmers although it would be lethal to the amebas. The bacterial flora and nutrients as well as the salinity, pH, temperature, and mineral or metal contents may influence the growth of virulent strains of free-living amebas. Therefore, the presence of free-living amebas in natural fresh water depends on multiple factors. The efforts to control the free-living amebic disease should be directed toward prevention.

REFERENCES

1. Abraham, S. N. and Lawande, R. V., Incidence of free-living amoebae in the nasal passages of local population in Zaria, Nigeria, *J. Trop. Med. Hyg.,* 85, 217, 1982.
2. Albújar, P. F., Meningoencefalitis por *Acanthamoeba* (Spanish), *Rev. Neuro-Psiquiatría (Lima, Peru),* 42, 137, 1979.
3. Anderson, K. and Jamieson, A., Primary amoebic meningoencephalitis, *Lancet,* 1, 902, 1972.
4. Anderson, K., Jamieson, A., Jadin, J. B., and Willaert, E., Primary amoebic meningoencephalitis, *Lancet,* 1, 672, 1973.
5. Anon., Primary amoebic meningoencephalitis in New Zealand, *N. Z. Med. J.,* 64, 164, 1969.
6. Apley, J., Clark, S. K. R., Roome, A. P., Sandry, S. A., Saygi, G., Silk, B., and Warhurst, D. C., Primary amoebic meningoencephalitis in Briain, *Br. Med. J.,* 1, 596, 1970.
7. Arce Vela, R. and Asato Higa, C., Encefalitis amebiana por *Acanthamoeba castellanii* (Spanish), *Diagnóstico 3 (Lima, Peru),* p. 25, 1979.
8. Ambrosino, M. and White, M., Producers, Bath Waters. Odyssey, BBC-TV Program, Public Broadcasting Associated, Newton, Mass., 1980.
9. Bedi, H. K., Devapura, J. C., and Bomb, B. S., Primary amoebic meningoencephalitis, *J. Ind. Med. Assoc.,* 58, 13, 1972.
10. Bhagwandeen, S. B., Carter, R. F., Naik, K. G., and Levitt, D., A case of *Hartmannellid* amebic meningoencephalitis in Zambia, *Am. J. Clin. Pathol.,* 63, 483, 1975.
11. Bhatia, P. S., Roy, S., and Ahuja, G. K., Meningoencephalitis due to soil ameba, *Neurology (India),* 27, 44, 1979.
12. Brass, K., Primare amoben meningoenzephalitis, *Dtsch. Med. Wochenschr.* 51, 1983, 1972.
13. Brass, K., Meningoencefalitis amebiásica primaria (por *Naeglerias*) (Spanish), *Arch. Venez. Med. Trop. Parasitol. Med.,* 5, 291, 1973.
14. Brown, T. J. and Cursons, R. T. M., Pathogenic free-living amebae (PFLA) from frozen swimming areas in Oslo, Norway, *Scand. J. Infect. Dis.,* 9, 237, 1977.
15. Brown, T. J., Cursons, R. T. M., and Keys, E. A., Amoebae from Antarctic soil and water, *Appl. Environ. Microbiol.,* 44, 491, 1982.
16. Brown, T. J., Disinfection against PAM amoebae (abstr.), *J. Protozool.,* 29, 496, 1982.
17. Brown, T. J., Cursons, R. T. M., Keys, E. A., Marks, M., and Miles, M,. The occurrence and distribution of pathogenic free-living amoebae (PFLA) in thermal areas of the North Island of New Zealand, *N.Z. J. Mar. Freshwater Res.,* 17, 59, 1983.
18. Campos, R., Gomes, de O., Prigenzi, L. S., and Stecca, J., Free-living amoebic meningoencephalitis: first case from Latin-America (Portuguese), *Rev. Inst. Med. Trop. São Paulo,* 19, 349, 1977.
19. Casemore, D. P., Contamination of virological tissue culture with a species of free-living soil amoeba, *J. Clin. Pathol.,* 22, 254, 1969.
20. Casemore, D. P., Free-living amoebae in home dialysis unit (correspondence), *Lancet,* 2, 1078, 1977.
21. Červa, L., Novák, K., and Culbertson, C. G., An outbreak of acute fatal amebic meningoencephalitis, *Am. J. Pathol.,* 88, 436, 1968.
22. Červa, L. and Novák, K., Amebic meningoencephalitis in Czechoslovakia. Preliminary report on the first 16 detected cases (in Czech), *Cesk. Epidemiol. Mikrobiol. Immunol. (Praha),* 17, 65, 1968.
23. Červa, L., Studies of *Limax* amoebae in a swimming pool, *Hydrobiologia (Hague),* 38, 141, 1971.
24. Červa, L., Tintěra, M., and Bendnář, J., Investigation of amoebae of the *Limax* group in a swimming pool during experimental application of iodine for water disinfection, *Czech. Hyg.,* 17, 389, 1972.

25. Červa, L., Serbus, C., and Skócil, V., Isolation of Limax amoebae from the nasal mucosa of man, *Folia Parasitol. (Praha)*, 20, 97, 1973.
26. Červa, L. and Huldt, G., Limax amoebae in five siwmming pools in Stockholm, *Folia Parasitol. (Praha)*, 21, 71, 1974.
27. Červa, L., The influence of temperature on the growth of *Naegleria fowleri* and *N. gruberi* in axenic culture, *Folia Parasitol. (Praha)*, 24, 221, 1977.
28. Červa, L., Jecna, P., and Hyhlik, R., *Naegleria fowleri* from a canal draining cooling water from a factory, *Folia Parasitol. (Praha)*, 27, 103, 1980.
29. Červa, L., Laboratory diagnosis of primary amoebic meningoencephalitis and methods for the detection of *Limax* amoebae in the environment, *Folia Parasitol. (Praha)*, 27, 1, 1980.
30. Červa, L. and Simanov, L., *Naegleria fowleri* in cooling circuits of industrial and power plants in North Moravia, *Folia Parasitol. (Praha)*, 30, 97, 1983.
31. Chang, R., Shiomi, T., and Franz, D., Temperature-induced cellular resistance to the "lipovirus", *Proc. Soc. Exp. Biol. (N.Y.)*, 115, 646, 1964.
32. Chang, S. L., Healy, G. R., McCabe, L., Shumaker, J. B., and Schultz, M. G., A strain of pathogenic *Naegleria* isolated from a human nasal swab, *Health Lab. Sci.*, 12, 1, 1975.
33. Chang, S. L., Resistance of pathogenic *Naegleria* to some common physical and chemical agents, *Appl. Environ. Microbiol.*, 35, 368, 1978.
34. Chinchilla, M., Castro, E., Alfaro, M., and Portilla, E., Free-living amoebae that can produce meningoencephalitis. First report in Costa Rica, *Rev. Latinoam. Microbiol.*, 21, 135, 1979.
35. Culbertson, C. G., Pathogenic free-living amebas, *Ind. Trop. Health*, 7, 118, 1970.
36. Culbertson, C. C., Smith, J. W., Cohen, H. K., and Minner, J. R., Experimental infection in mice and monkeys by *Acanthamoeba*, *Am. J. Pathol.*, 35, 185, 1959.
37. Cursons, R. T. M. and Brown, T. J., The 1968 cases of primary amoebic meningoencephalitis — Myxomycete or *Naegleria?*, *N.Z. Med. J.*, 82, 123, 1975.
38. Cursons, R. T. M. and Brown, T. J., Identification and classification of the aetiological agents of primary amebic meningoencephalitis, *N.Z. J. Mar. Freshwater Res.*, 10, 245, 1976.
39. Cursons, R. T. M., Brown, T. J., Burns, B. J., and Taylor, D. E. M., Primary amoebic meningoencephalitis contracted in a thermal tributary of the Waikato River-Taupo: a case report, *N.Z. Med. J.*, 84, 479, 1976.
40. Cursons, R. T. M., Brown, T. J., and Culbertson, C. G., Immunoperoxidase staining of trophozoites in primary amoebic meningoencephalitis, *Lancet*, 2, 479, 1976.
41. Cursons, R. T. M., Brown, T. J., and Keys, E. A., Primary amoebic meningoencephalitis (PAM) in New Zealand—aetiological agents, distribution, occurrence, and control, *Proc. 9th N.Z. Biotech. Conf.*, p. 96, 1977.
42. Cursons, R. T. M., Brown, T. J., and Keys, E. A., Immunity to pathogenic free-living amoeba, *Lancet*, 2, 875, 1977.
43. Cursons, R. T. M., Brown, T. J., Keys, E. A., Gordon, E. H., Leng, R. H., Havill, J. H., and Hyne, B. E., Primary amoebic meningoencephalitis in an indoor heat exchange swimming pool, *N.Z. Med. J.*, 89, 330, 1979.
44. Cursons, R. T. M., Brown, T. J., and Keys, E. A., Effect of disinfectants on pathogenic free-living amoebae: in axenic conditions, *Appl. Environ. Microbiol.*, 40, 62, 1980.
45. Das, S. R., Isolation of *Naegleria* and *Hartmannella* amoebae from Beckenham (London) soils and their pathogenicity in mice, *Trans. R. Soc. Trop. Med. Hyg.*, 66, 663, 1972.
46. Davis, P. G., Caron, D. A., and Sieburth, J. McN., Oceanic amoebae from the North Atlantic: culture, distribution, and taxonomy, *Trans. Am. Microsc. Soc.*, 97, 73, 1978.
47. Dawson, M. W., Brown, T. J., and Till, D. G., The effect of Baquacil on pathogenic free-living amoebae (PFLA). I. In axenic conditions, *N.Z. J. Mar. Freshwater Res.*, 17, 305, 1983.
48. Dawson, M. W., Brown, T. J., Biddick, C. J., and Till, D. G., The effect of Baquacil on pathogenic free-living amoebae (PFLA). II. In simulated natural conditions — in the presence of bacteria and/or organic matter, *N.Z. J. Mar. Freshwater Res.*, 17, 313, 1983.
49. De Jonckheere, J., Van Dijck, P., and Van de Voorde, H., The effects of thermal pollution of the distribution of *Naegleria fowleri*, *J. Hyg. (Cambridge)*, 75, 7, 1975.
50. De Jonckheere, J. and Van de Voorde, H., Differences in destruction of cysts of pathogenic and nonpathogenic *Naegleria* and *Acanthamoeba* by chlorine, *Appl. Environ. Microbiol.*, 31, 294, 1976.
51. De Jonckheere, J. and Van de Voorde, H., The distribution of *Naegleria fowleri* in man-made thermal waters, *Am. J. Trop. Med. Hyg.*, 26, 10, 1977.
52. De Jonckheere, J., Use of an axenic medium for differentiation between pathogenic and non pathogenic *Naegleria fowleri* isolates, *Appl. Environ. Microbiol.*, 33, 751, 1977.
53. De Jonckheere, J., Quantitative study of *Naegleria fowleri* in surface water, *Protistologica*, 14, 475, 1978.
54. De Jonckheere, J., Occurrence of *Naegleria* and *Acanthamoeba* in Aquaria, *Appl. Environ. Microbiol.*, 38, 590, 1979.

55. De Jonckheere, J., Studies on pathogenic free-living amoebae in swimming pools, *Bull. Inst. Pasteur*, 77, 385, 1979.

56. De Jonckheere, J., Pathogenic free-living amoebae in swimming pools: survey in Belgium, *Ann. Microbiol. (Inst. Pasteur)*, 130B, 205, 1979.

57. De Jonckheere, J. F., Growth characteristics, cytopathic effect in cell culture, and virulence in mice of 36 type strains belonging to 19 different *Acanthamoeba* spp., *Appl. Environ. Microbiol.*, 39, 681, 1980.

58. De Jonckheere, J. F., Pathogenic and nonpathogenic *Acanthamoeba* spp. in thermally polluted discharges and surface waters, *J. Protozool.*, 28, 56, 1981.

59. De Jonckheere, J. F., Hospital hydrotherapy pools treated with ultraviolet light: bad bacteriological quality and presence of thermophilic *Naegleria*, *J. Hyg. (Cambridge)*, 88, 205, 1982.

60. De Jonckheere, J. F., Melard, C., and Philippart, J. C., Appearance of pathogenic *Naegleria fowleri, Amoebida, Vahlkampfiidae)* in artificially heated water of a fish farm, *Aquaculture*, 35, 73, 1983.

61. Dillon, R. D., The ecology of free-living and parasitic protozoa of Antarctica, *Antarct. J.*, 104, 1967.

62. Dive, D. G., Leclerc, H., DeJonckheere, J., and DeLatte, J. M., Isolation of *Naegleria fowleri* from the cooling pond of an electric power plant in France, *Ann. Microbiol. (Inst. Pasteur) (Paris,)* 132 A, 97, 1981.

63. Dorsch, M. M., Cameron, A. S., and Robinson, B. S., The epidemiology and control of primary amoebic meningoencephalitis with particular reference to South Australia, *Trans. R. Soc. Trop. Med. Hyg.*, 77, 372, 1983.

64. Duma, R. J., Shumaker, J. B., and Callicott, J. H., Primary amebic meningoencephalitis. A survey in Virginia, *Arch. Environ. Health*, 23, 43, 1971.

65. Dvořák, R. and Sckočil, V., Amoeba of the *Limax* group in the nasal mucous membrane (in Czech), *Cesk. Otolaryngol. (Praha)*, 21, 279, 1972.

66. Edwards, J. H., Griffiths, A. J., and Mullins, J., Protozoa as sources of antigen in "humidifier fever" (correspondence), *Nature (London)*, 264, 438, 1976.

67. Gottlieb, B. and Reyes, H., Free-living amoebae: new pathogenic agents for human beings, *Rev. Med. Chile (Santiago)*, 105, 467, 1977.

68. Griffin, J. L., Temperature tolerance of pathogenic and nonpathogenic free-living amoebas, *Science*, 178, 869, 1972.

69. Griffin, J. L., Environmental sampling for pathogenic *Naegleria (abstr.)*, *J. Protozool.*, 20 (Suppl.), p. 497, 1973.

70. Grillot, R. and Ambroise-Thomas, P., Free-living amoebae in swimming pools in the Grenoble area. Influence of "summer-winter" use and of sterilization method, *Rev. Epidemiol. Santé Publique*, 28, 185, 1980.

71. Guevara-Benitez, D., Mascaró-Lazcano, C., and Fluvia-Bru, C., The presence of amoebae of the Limax group in various collections of fresh water, *Rev. Ibérica Parasitol.*, 38, 615, 1978.

72. Hewitt, R., The natural habitat and distribution of *Hartmannella castellanii* (Douglas). A reported contaminant of bacterial cultures, *J. Parasitol.*, 23, 491, 1937.

73. Holbrook, T. W., Boackle, R. J., Parker, B. W., and Vesely, J., Activation of the alternative complement pathway by *Naegleria fowleri*, *Infect. Immun.*, 30, 58, 1980.

74. Hull, R., Minner, J., and Mascoli, C., New viral agents recovered from tissue cultures of monkey kidney cells. III. Recovery of additional agents both from monkey tissues and directly from tissues and excreta, *Am. J. Hyg.*, 68, 31, 1958.

75. Jacquemin, J. L., Jacquemin, P., Flohic, A. M, and Olory-Togbe, P., Présence d'amibes de type Limax dans les eaux de distribution de L'agglomération de Rennes, *Bull. Soc. Pathol. Exotique*, 65, 253, 1974.

76. Jacquemin, J. L., Simitzis-Le Flohic, A. M., and Chauveau, N., Free-living amoebae in fresh water. A study of the water supply of the town of Poitiers, *Bull. Soc. Pathol. Exotique*, 74, 521, 1981.

77. Jadin, J. B., Willaert, E., and Compere, F., De la necessité du controle biologique des eaux potables, *Bull. Acad. Natl. Med. (Paris)*, 156, 995, 1972.

78. Jadin, J. B. and Willaert, E., Au sujet de la dispersion des amibes du groupe Limax, *Protistologica*, 8, 505, 1972.

79. Jadin, J. B., Willaert, E., and Hermanne, J., Presence d'amibes limax dans l'intestin de l'homme et des animaux, *Arsom Bull. Seances*, 3, 520, 1973.

80. Jadin, J. B,. Hypothèses au sujet de l'adaptation des amibes du groupe limax à l'homme et aux animaux, *Ann. Parasitol. Hum. Comp.*, 48, 199, 1973.

81. Jadin, J. B., De la dispersion et du cycle des amibes libres, *Ann. Soc. Belge Med. Trop.*, 54, 371, 1974.

82. Jadin, J. B., Amibes ("limax") vecteurs possibles de mycobácteries et de *Mycobacterium leprae*, *Acta Leprol.*, 59—60, 57, 1975.

83. Jamieson, A. and Anderson, K., A method for the isolation of *Naegleria* species from water samples, *Pathology*, 5, 55, 1973.

84. Janitschke, K., Werner, H., and Muller, G., Das vorkommen von fieilebenden amöben mit moglichen pathogenen eigenschaften in schwimmbadern, *Zentralbl. Bakteriol. Orig. B. Hyg.*, 170, 108, 1980.

85. Janitschke, K., Lichy, S., and Westphal, C., Examination of thermally polluted water for free-living amoebae and testing for their possible pathogenic properties, *Zentralbl. Bakteriol. Orig. B. Hyg.*, p. 160, 1982.

86. Janitschke, K., Lichy, S., and Thalmann, U., Experiments on the pathogenicity of *Acanthamoeba* isolated from the environment, *Zentralbl. Bakteriol. Orig. B. Hyg.*, 177, 350, 1983.

87. Jírovec, O. and Kneiflová-Jírovcová, J., La résistance des amibes du type limax vers quelques facteurs externes, *J. Parasitol.*, 56, 172, 1970.

88. Kadlec, V., The effects of some factors on the growth and morphology of *Naegleria* sp. and three strains of the genus *Acanthamoeba, Folia Parasitol. (Praha)*, 22, 317, 1975.

89. Kadlec, V., Skvarova, J., Červa, L., and Nebazniva, D., Virulent *Naegleria fowleri* in indoor swimming pool, *Science*, 201, 1025, 1978.

90. Kadlec, V., Skvarova, J., Červa, L., and Nebazniva, D., Virulent *Naegleria fowleri* in indoor swimming pool, *Folia Parasitol. (Praha)*, 27, 11, 1980.

91. Kadlec, V., Different virulence of *Naegleria fowleri* strains isolated from a swimming pool, *Folia Parasitol. (Praha)*, 28, 97, 1981.

92. Kasprzak, W., Primary amebic meningoencephalitis caused by *Naegleria* sp. A new problem in medical parasitology (in Polish), *Wiadomosci Parazytol.*, 17, 273, 1971.

93. Kasprzak, W. and Mazur, T., Free-living amoebae isolated from waters frequented by people in the vicinity of Poznan, Poland. Experimental studies in mice on the pathogenicity of the isolates, *Z. Tropenmed. Parasitol.*, 23, 391, 1972.

94. Kasprzak, W. and Mazur, T., Method of isolation of free-living amoebae from their natural habitat (in Polish), *Wiadomosci Parazytol.*, 19, 855, 1973.

95. Kasprzak, W., Mazur, T., and Rucka, A., Studies on some pathogenic strains of free-living amoeba isolated from lakes in Poland, *Ann. Soc. Belge Med. Trop.*, 54, 351, 1974.

96. Kasprzak, W. and Mazur, T., The effect of thermic pollution of waters on the distribution of pathogenic *Naegleria* Strains (in Polish), *Wiadomosci Parazytol.*, 42, 457, 1976.

97. Kingston, D. and Warhurst, D. C., Isolation of amoeba from the air, *J. Med. Microbiol.*, 2, 27, 1969.

98. Koski, T. A., Stuart, L. S., and Ortenzio, L. F., Comparison of chlorine, bromine, and iodine as disinfectants for swimming pool water, *Appl. Microbiol.*, 14, 276, 1966.

99. Lamy, L. and Fromentin, H., Interet particulier de la culture pour la mise en evidence des amibes libres susceptibles de s'installer chez l'homme et de devenir pathogene, *C. R. Acad. Sci. (Paris) Ser. D*, 277, 1205, 1973.

100. Lastovica, A. J. and Elsdon-Dew, R., Primary amoebic meningoencephalitis caused by *Naegleria* species, *S. Afr. J. Sci.*, 67, 464, 1971.

101. Lastovica, A. J., Isolation, distribution and disease potential of *Naegleria* and *Acanthamoeba* (order: *Amoebida*) in South Africa, *Trans. R. Soc. S. Afr.*, 44, 269, 1980.

102. Lawande, R.V., Abraham, S. N., John, I., and Egler, L. J., Recovery of soil amebas from nasal passages of children during the dusty Harmattan period in Zaria, *Am. J. Clin. Pathol.*, 71, 201, 1979.

103. Lawande, R. V., Ogukanmi, A. E., and Egler, L. J., Prevalence of pathogenic free-living amoebae in Zaria, Nigeria, *Ann. Trop. Med. Parasitol.*, 73, 51, 1979.

104. Lyons, T. B. and Kapur, R., Limax amoebae in public swimming pools of Albany, Schenectady, and Rensselaer Counties, New York: their concentration, correlations, and significance, *Appl. Environ. Microbiol.*, 33, 551, 1977.

105. Madrid, L. R., Meningoencefalitis amébica primaria (MAP), *Rev. Med. Hondurena (Spanish)*, 50, 200, 1983.

106. Madrigal-Sesma, M. J., Santillana, I., and Zapatero Ramos, L. M., Presencia de amebas limax en aguas naturales de Madrid (Spanish), *Rev. Iber. Parasitol.*, Extra, 125, 1982.

107. Mascaró, C., Fluvía, C., Osuna, A., and Guevara, D., Virulent *Naegleria* sp. isolated from a river in Cadiz (Spain), *J. Parasitol.*, 67, 599, 1981.

108. Mascaró-Lazcano,M. C., Guevara-Benítez, D., and de la Rubia-Nieto, A., Aspectos metodológicos en el estudio de "amebas limax", *Rev. Ibérica Parasitol.*, 37, 329, 1977.

109. Michel, R.and Schneider, H., Untersuchungen zum vorkommen von limax-amöben im therapie schwimmbad eines krankenhauses, *Zentralbl. Bakteriol. Orig. B. Hyg.*, 170, 479, 1980.

110. Michel, R., Röhl, R., and Schneider, H., Isolierung von freilebenden amöben durch Gewinnung von Nasenschleimhautabstrichen bei gesunden Probanden, *Zentralbl. Bakteriol. Orig. B. Hyg.*, 176, 155, 1982.

111. Michel, R. and De Jonckheere, J., Erster nachweis einer pathogenen *Naegleria* — Art *(N. australiensis* De Jonckheere (1981) in Deutschland, *Z. Parasitenkunde*, 69, 395, 1983.

112. Michel, R. and DeJonckheere, J. F., Isolation and identification of pathogenic *Naegleria australiensis* (De Jonckheere, 1981) from pond water in India (letter), *Trans. R. Soc. Trop. Med. Hyg.*, 77, 878, 1983.

113. Miller, G., Cullity,G., Walpole, I., O'Connor, J., and Masters, P., Primary amoebic meningoencephalitis in Western Australia, *Med. J. Aust., P.*, 352, 1982.

114. Molet, B., Derr-Harf, C., Schreiber, J. E., and Kremer, M., Étude des amibes libres dans les eaux de Strasbourg, *Ann. Parasitol. Hum. Comparée*, 51, 401, 1976.

115. Molet, B., Feki, A., Haag, R., and Kermer, M., Isolement d'amibes libres par ecouvillonnage nasal pratique chez 300 sujets sains, *Rev. Oto-Neuro-Ophtalmol. (Paris)*, 53, 121, 1981.

116. Nakamura, N., On a strain of amoeba accidentally discovered on agar medium which is phagocytic to bacteria, *Kitasato Arch. Exp. Med.*, 24, 23, 1951.

117. Nakanishi, K., A new type ameba *(Amoeba ferox)* phagocyting pathogenic intestinal bacteria, recovered from river water in Java, *Jpn. Med. J.*, 3, 231, 1950.

118. Napolitano, J. J., Isolation of amoebae from edible mushrooms, *Appl. Environ. Microbiol.*, 44, 255, 1982.

119. Newsome, A. L. and Wilhelm, W. E., Inhibition of *Naegleria fowleri* by microbial iron-chelating agents: ecological implications, *Appl. Environ. Microbiol.*, 45, 655, 1983.

120. Newsome, A. L. and Wilhelm, W. E., Effect of exogenous iron on the viability of pathogenic *Naegleria fowleri* in serum, *Experientia*, 37, 1160, 1983.

121. Nicoll, A. M., Fatal primary amoebic meningoencephalitis, *N.Z. Med. J.*, 78, 108, 1973.

122. O'Dell, W. D., Isolation, enumeration and identification of amoebae from a Nebraska Lake, *J. Protozool.*, 26, 265, 1979.

123. Pennisi, L., Mento, G., and Todaro, F., Sulla diffusione di anticorpi anti-*Acanthamoeba castellanii* in soggetti provenienti da varie regione Italiana, *Parassitologia*, 13, 299, 1971.

124. Pernin, P. and Riany, A., Étude sur la presence d'amibes libres dans les eaux des piscines Lyonnaises, *Ann. Parasitol. (Paris)*, 53, 333, 1978.

125. Pernin, P., Riany, A., and Grimaud, J. A., Ultrastructural study of experimental meningoencephalitis induced by a strain of *Acanthamoeba* isolated from a swimming pool, *Protistologica*, 15, 307, 1979.

126. Proca-Ciobanu, M. I., Lupascu, G. H., and Steriu, D., Isolation of a pathogenic strain *A. castellanii* in Roumania, *Arch. Roum. Pathol. Exp.*, 32, 205, 1973.

127. Rivera, F., Galván, M., Robles, E., Leal, P., González, L., and Lacy, A. M., Bottled mineral waters polluted by protozoa in México, *J. Protozool.*, 28, 54, 1981.

128. Rowbotham, T. J, Preliminary report on the pathogenicity of Legionella pneumophila for freshwater and soil amoebae, *J. Clin. Pathol.*, 33, 1179, 1980.

129. Rowbotham, T. J., Pontiac fever explained?, *Lancet, p.* 969, 1980.

130. Salazar, H. C., Moura, H., and Ramos, R. T., Isolamento de amebas de vida livre a partir de agua mineral engarrafada, *Rev. Saude Publica (Sao Paulo)*, 16, 261, 1982.

131. Sawyer, T. K., Visvesvara, G. S., and Harke, B. A., Pathogenic amoebas from brackish and ocean sediments with a description of *Acanthamoeba hatchetti*, n.sp., *Science*, 196, 1324, 1977.

132. Scaglia, M., Strosselli, M., Grazioli, V., Gatti, S., Capelli, D., and Bernuzzi, A. M., Free-living amoebae, agents of meningoencephalitis in man: epidemiological investigations in non potable waters in the town of Pavia, *G. Malattie Infettive Parassitarie*, 34, 1140, 1982.

133. Scaglia, M., Strosselli, M., Grazioli, V., and Gatti, S., Pathogenic *Naegleria:* isolation from thermal mud samples in a northern Italian spa, *Trans. R. Soc. Trop. Med. Hyg.*, 77, 136, 1983.

134. Scholten, T., Soil amoebae in Canada (corespondnce *Can. Med. Assoc. J.*, 120, 267, 1979.

135. Shookhoff, H. B., Meningoencephalitis due to free-living amoebas normally found in soil, *Ann. Intern. Med.*, 70, 1276, 1969.

136. Shumaker, J. B., Healy, G. R., English, D., Schulz, M., and Page, F., *Naegleria gruberi* isolation from nasal swab of a healthy individual, Lancet, 2, 602, 1971.

137. Simitzis, A.-M., LeGoff, and L'Azou, M.-T., Isolement d'amibes libres a partir de la muqueuse nasale de l'homme. Risque eventuel (Memoires Originaux), *Ann. Parasitol. Hum. Comparée*, 54, 121, 1979,

138. Singh, B. N. and Das, S. R., Intranasal infection of mice with flagellate stage of *Naegleria aerobia* and its bearing on the epidemiology of human meningoencephalitis, *Curr. Sci. (India)*, 41, 625, 1972.

139. Singh, B. N. and Das, S. R., Occurrence of pathogenic *Naegleria aerobia, H. culbertsoni* and *H. rhysodes* in sewage sludge samples of Lucknow, *Curr. Sci. (India)*, 41, 277, 1972.

140. Singh, B. N. and Hanumaiah, V., Temperature tolerance of free-living amoeba and their pathogenicity to mice, *Ind. J. Parasitol.*, 1, 71, 1977.

141. Skočil, V., Červa, L., and Serbus, C., Epidemiological study of amoebas of the Limax group in military communities. First reports, *J. Hyg. Epidemiol. Microbiol. Immunol. (Praha)*, 14, 61, 1970.

142. Skočil, V., Červa, L., Serbus, C., and Nejedlo, V., Epidemiological study of amoebas of the Limax group in military communities. II. Study of the military community L. 1 (1968—1969), *J. Hyg. Epidemiol. Microbiol. Immunol. (Praha),* 14, 324, 1970.

143. Skočil, V., Serbus, C., and Červa, L., Epidemiological study of the incidence of amoebas of the Limax group in military communities. III. Investigation of the community of 3rd Garrison L1— problems of contagion in the community, *J. Hyg. Epidemiol. Microbiol. Immunol. (Praha),* 15, 156, 1971.

144. Skočil, V., Červa, L., and Serbus, C., Epidemiological study of amoeba of the Limax group in military communities. Relation between the findings of amoeba in the external environment and their incidence in the soldiers during the investigation into the community L. IV, *J. Hyg. Epidemiol. Microbiol. Immunol. (Praha),* 15, 445, 1971.

145. Skočil, V., Dvorak, R., Sterba, J., Slăjs, J., Serbus, C., and Červa, L., Epidemiological study of the incidence of amoebas of the Limax group in military communities. V. Relation between the presence of amoebas of the Limax group in nasal swabs and a pathological finding in nasal mucosa, *J. Hyg. Epidemiol. Microbiol. Immunol. (Praha),* 16, 101, 1972.

146. Skočil, V., Serbus, C., and Červa, L., Epidemiological study of the incidence of the Limax group in military communities. VI. Relation between the finding of amoebas of the Limax group in nasal swabs and some epidemiological indices, amoebas of the Limax group in military, *J. Hyg. Epidemiol. Microbiol. Immunol. (Praha),* 16, 226, 1972.

147. Soh, C. T., Inn, K., Chang, B. P., and Hwang, H. K., Pathogenicity of free-living amoebae isolated from various places in Seoul, *5th Int. Congr. Protozool.,* P. 412, 1977.

148. Stevens, A. R., Tyndall, R. L., Coutant, C. C., and Willaert, E., Isolation of the etiological agent of primary amebic meningoencephalitis from artificially heated waters, *Appl. Environ. Microbiol.,* 34, 701, 1977.

149. Sykora, J., Karol, M., Keleti, G., and Novak, D., Amoebae as sources of hypersensitivity pneumonitis, *Environ. Int.,* 8, 343, 1982.

150. Sykora, J. L., Keleti, G., and Martinez, A. J., Occurrence and pathogenicity of *Naegleria fowleri* in artificially heated waters, *Appl. Environ. Microbiol.,* 45, 974, 1983.

151. Symmers, W. St.C, Primary amoebic meningoencephalitis in Britain, *Br. Med. J.,* 4, 449, 1969.

152. Umeche, N., The numbers of *Naegleria* spp. in Michigan soil and litters, *Arch. Protistenkunde,* 127, 127, 1983.

153. Van den Driessche, E., Vandepitte, J., Van Dijck, P. J., De Jonckheere, J., and Van de Voorde, H., Primary amebic meningoencephalitis after swimming in stream water, *Lancet,* 2, 971, 1973.

154. Villacorta, B. F. and Jueco, N. L., Pathogenicity of *Acanthamoeba* on mice, *Kalikasan Philipp. J. Biol.,* 10, 337, 1981.

155. Visvesvara, G. S., Brandt, F. H., Baxter, P. J., and Healy, G. R. Isolation of a pathogenic *Acanthamoeba polyphaga* from disposable filters attached to heating, ventilation and air conditioning (HVAC) units and demonstration of anti-*Acanthamoeba* antibody in human sera, *J. Protozool.,* (29, (Abstr), 489, 1982.

156. Visvesvara, G. S., Mirra, S. S., Brandt, F. H., Moss, D. M., Mathews, H. M., and Martinez, A. J., Isolation of two strains of *Acanthamoeba castellanii* from human tissue and a note on their pathogenicity and isoenzyme profiles, *J. Clin. Microbiol.,* 18, 1405, 1983.

157. Wang, S. S. and Feldman, H. A., Occurrence of *Acanthamoeba* in tissue cultures inoculated with human pharyngeal swabs, *Antimicrob. Agents Chemother.,* 1, 50, 1961.

158. Wang, S. S. and Feldman, H. A., Isolation of *Hartmannella* species from human throats, *N. Engl. J. Med.,* 277, 1174, 1967.

159. Warhurst, D. and Armstrong, J., A study of a small amoeba from mammalian cell cultures infected with Ryan virus, *J. Gen. Microbiol.,* 50, 207, 1968.

160. Warhurst, D. C., Carman, J. A., and Mann, P. G., Survival of *Naegleria fowleri* cysts at 4°C for eight months with retention of virulence, *Trans. R. Soc. Trop. Med. Hyg.,* 74, 832, 1980.

161. Wellings, F. M., Amuso, P. T., Chang, S. L., and Lewis, A. L., Isolation and identification of pathogenic *Naegleria* from Florida lakes, *Appl. Environ. Microbiol.,* 34, 661, 1977.

162. Weng, N. K., Wagner, W., and Parker, J. C., Primary amebic meningoencephalitis a potential problem in the Southeastern United States, *South. Med. J. (Birmingham, Ala.),* 64, 691, 1971.

163. Willaert, E., Jamieson, A., Jadin, J. B., and Anderson, K., Epidemiological and immunoelectrophoretic studies on human and environmental strains of *Naegleria fowleri, Ann. Soc. Belge Med. Trop.,* 54, 333, 1974.

164. Willaert, E. and Stevens, A. R. Isolation of pathogenic amoeba from thermal-discharge water, *Lancet,* 1, 741, 1976.

Chapter 5

THE DISEASE: CLINICAL TYPES — MANIFESTATIONS AND COURSE

I. ACUTE INFECTION DUE TO *NAEGLERIA FOWLERI*

Central nervous system (CNS) disease due to free-living or amphizoic amebas includes Primary Amebic Meningoencephalitis (PAM) due to *Naegleria fowleri* and Granulomatous Amebic Encephalitis (GAE) due to *Acanthamoeba* spp.[1,2,6,8-10,25,26,52,54-57,63-65,67,72,74,78-81,84,88,92,93,101,102,111,114,115]

PAM usually occurs in children and young adults who have previously been in good health. Victims usually have had a recent history of swimming in heated swimming pools, man-made lakes, or had some contact with water or mud.[10,73] However, cases have been reported where there was no contact with water.[26,27,35]

Subclinical infections due to free-living amebas are possible in healthy individuals with the protozoa living as "normal flora" in the nose and throat. It is also possible that antibodies and cell-mediated immunity protect the host in ordinary circumstances against infection.[46,48] Thus far, only *N. fowleri* has been implicated in human PAM. Other species such as *N. australiensis, N. lovaniensis,* and *N. jadini* have not been found to produce spontaneous disease in humans or animals.

A. Pathogenesis

The olfactory neuroepithelium is the route of invasion in PAM due to *N. fowleri.*[16,45,70]

Destruction of the olfactory mucosa and olfactory bulbs and hemorrhagic necrosis of both gray and white matter with an inflammatory infiltrate consisting of abundant polymorphonuclear leukocytes, eosinophils, and few macrophages are the histopathological characteristics. Only trophozoites are found in the lesions.[34]

B. Portal of Entry

The olfactory neuroepithelium is the anatomical site of the primary lesion in PAM due to *N. fowleri.*[69] Respiratory symptoms may be the result of subclinical *N. fowleri* or *Acanthamoeba* spp. infection or hypersensitivity. The pathogenic amebas probably enter the nasal cavity by inhalation or aspiration of water containing the trophozoites or cysts. Inhalation or aspiration by aerosols containing the cysts is another possible source.

C. Incubation Period

The period between the initial contact with the protozoon and the onset of clinical signs of the disease (fever, headache, rhinitis) is the incubation period. This can be a variable length of time. In most cases, it is from 2 to 3 days, although periods as long as 7 to 15 days have been noted.

D. Signs and Symptoms

PAM is a disease with an abrupt onset and fulminant course. The symptoms are primarily those associated with severe meningeal irritation and consist of severe headache, stiff neck, fever (39 to 40°C), and vomiting. Pharyngitis or symptoms of nasal obstruction and discharge are less frequent (Table 1). An occasional complaint in the first day or so is the distortion of taste or smell. Headache, vomiting, and fever persist, but within 2 to 4 days after onset, drowsiness, confusion, and neck stiffness develop. Convulsions may occur but have not been pronounced in most cases. Progressive de-

Table 1
PRESENTING SYMPTOMS AND SIGNS OF 15
PATIENTS WITH AUTOPSY-PROVEN PAM
DUE TO *NAEGLERIA FOWLERI*

Clinical symptoms/signs	No. of patients (%)
Headache	15 (100)
Anorexia, nausea, vomiting	14 (93)
Fever (39—40°C)	14 (93)
Meningism	13 (86)
Mental status abnormalities and behavioral changes[a]	10 (66)
Coma (as a presenting symptom or shortly after admission)	4 (26)
Visual disturbances; diplopia; blurred vision	2 (13)
Seizures	2 (13)
Parosmia	2 (13)
Ageusia	2 (13)
Diabetes insipidus (agonal)	1 (6)

[a]   Lethargy, somnolence, drowsiness, obtundation, confusion, irritability, restlessness, hallucinations.

terioration follows, leading to deep coma with minimal if any focal neurologic signs. The vast majority of cases have ended fatally in about 1 week after the appearance of the first symptoms.[5,50,51,73] The premortem diagnosis is established by finding trophozoites in the cerebrospinal fluid (CSF) (see Chapter 7, Table 1). Cardiac abnormalities in PAM are common.[68]

The peripheral white blood cell count is almost constantly elevated with a marked increase in polymorphonuclear leukocytes. Some variability in the CSF findings has occurred, but in most reports the fluid has been purulent with a predominance of neutrophils and, therefore, a bacterial infection was suspected. In a few instances, the initial CSF early in the illness contains clear fluid and a number of erythrocytes with a modest leukocytic response. The findings might be confused with *Herpesvirus hominis* or bacterial encephalitis.[112] The protein content is elevated in almost all patients with PAM, with a range between 100 and 1000 mg/100 mℓ. The glucose concentration has been mildly reduced in some cases and normal in others. Motile amebas can be visualized in the CSF with wet preparations (see Chapter 7, Figures 1 and 2).

E. Differential Diagnosis

No distinctive differences exist which allows differentiation of PAM from acute pyogenic or bacterial meningoencephalitis on clinical grounds. A history of previous good health and swimming in fresh water shortly before the onset of illness are helpful in making a diagnosis of PAM.

Computed tomographic studies (CT scans) show obliteration of the cisterns surrounding the midbrain and the subarachnoid space over the hemispheric convexities (Figures 1A and 1B). Marked enhancement in these regions may be seen after administration of intravenous contrast medium (Figures 1C and 5D).[62]

F. Prognosis

Serological studies[40-42] in free-living amebas have been performed and reported.[21-24,40,42] They have led to significant accumulation of data on the immune response of the host. In addition to great conceptual potential, these studies also have considerable practical implication for diagnosis, treatment, and possible prevention of the disease.[44,89,91,96,99,105,107,109]

The role of cell-mediated immunity using the in vitro macrophage inhibition test and

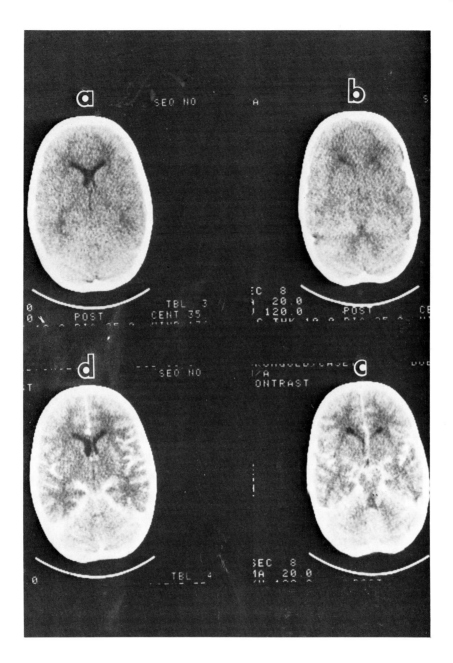

FIGURE 1. (a) and (b) Precontrast CT scans of the head of a case of PAM show obliteration of the ambiens interpeduncular and quadrigeminal cisterns. (c) and (d) Postcontrast CT scans of a case of PAM of the corresponding levels showing marked enhancement of the obliterated cisterns as well as sulci and adjacent gray matter. (From Lam, A. H., deSilva, M., Procopis, P., and Kan, A., *J. Comput. Assist. Tomogr.*, 6, 620, 1982. With permission.)

the in vivo delayed hypersensitivity test suggests responses to both heterologous and homologous antigens. This indicates that exposure to free-living amebas stimulates the immune system.[24,91]

II. CHRONIC "GRANULOMATOUS" INFECTION DUE TO *ACANTHAMOEBA* SPP.

Granulomatous amebic encephalitis (GAE) refers to an illness with distinct clinico-pathological characteristics usually occurring in debilitated and chronically ill individuals, some of whom have undergone immunosuppressive therapy. GAE is usually manifest by focal neurological deficits, signs of increased intracranial pressure, and neuroradiographic features suggestive of an expanding intracranial mass. It is important to emphasize the focal or multifocal nature of the CNS lesions in GAE. The appearance of focal neurologic signs, even though they are subtle, and the radiologic demonstration of space occupying avascular masses with vascular or ventricular shifts in a patient with encephalitis should alert the physician to the possibility of GAE. The possibility of focal expanding lesions may be further supported by the results of ancillary studies such as EEG, CT scan, and angiography.

Cases of GAE due to *Acanthamoeba* spp. *(A. castellanii* and *A. culbertsoni* and, possibly, *Vahlkampfia)* have been reported in chronically ill and debilitated[38,49] or immunologically impaired individuals,[113] some of whom were iatrogenically immunosuppressed.[47] However, some do not demonstrate an immunodeficiency.[85,86,97] In debilitated and chronically ill individuals with depressed or suppressed cell-mediated immunity and decreased antibody formation, protozoa may proliferate, act as "opportunistic" organisms, and cause a fulminant infection.[71]

In the future, it will be important to evaluate the functional and morphological features of the thymus, lymph nodes and spleen in order to evaluate the possibility of impairment of cell-mediated immune response or other immune deficit.

Viral, bacterial, and fungal infections of the skin have "significantly increased" in incidence and severity in immunosuppressed patients. Patients involved in organ transplantation or under treatment with antineoplastic drugs, broad spectrum antibiotics, parenteral hyperalimentation, specialized apparatus such as ventilation equipment or renal dialysis, and recipients of immunosuppressive drug therapy have increased susceptibility to infection. Other important factors include malignant diseases (especially lymphoma), chronic illnesses, drug addiction, diabetes mellitus, trauma, burns, and geographic and occupational considerations.

The term opportunistic is probably meaningless in a general sense since almost any microbial organism can produce severe disease if the conditions are appropriate. Almost any pathogen can live in peaceful symbiosis in a disease-free host. Infection with *Acanthamoeba* spp. may result in a wide spectrum of diseases. On one side there is GAE with the characteristic necrosis of CNS tissue, chronic inflammation, and multinucleated foreign body giant cell reaction. On the other side, there is the indolent form of the disease characterized by necrotizing encephalitis with minimal subacute inflammatory response and without granulomatous reaction.

A. Pathogenesis

The route of invasion and penetration into the CNS in cases of GAE appears to be hematogenous, probably from a primary focus in the skin or lower respiratory tract. Chronic ulceration of the skin containing amebic trophozoites and cysts has been reported in patients that died of GAE.[49] Purulent discharge from the ears[58] or mouth and ulcerations of the ocular tissues have also been reported.[76] The virulent *Acanthamoeba* may enter the respiratory tract by inhalation of air containing cysts or aerosols. Person-to-person transmission can probably be excluded.

B. Portal of Entry

The portal of entry of *Acanthamoeba* spp. into the body may be the lower respiratory tract, ulcerations of the skin or mucosa, or other open wounds.

Table 2

PRESENTING SYMPTOMS AND SIGNS OF 15 PATIENTS
WITH AUTOPSY/BIOPSY-PROVEN CASES OF AMEBIC
ENCEPHALITIS DUE TO *ACANTHAMOEBA* spp.

Clinical symptoms/signs	No. of patients (%)
Mental status abnormalities and behavioral changes[a]	13 (86)
Seizures	10 (66)
Fever (39—40°C)	8 (53)
Hemiparesis	8 (53)
Headache	8 (53)
Meningism	6 (40)
Cranial nerve palsies; diplopia and other visual disturbances	4 (26)
Ataxia	3 (20)
Aphasia	3 (20)
Anorexia; nausea; vomiting	3 (20)
Papilledema	1 (6)

[a] Irritability, confusion, hallucinations, dizziness, somnolence, drowsiness.

C. Incubation Period

The incubation period of GAE is unknown. Several weeks or months are necessary to establish the disease. The clinical course is prolonged. The clinical, radiological, and pathological data are not helpful in determining how long the amebic organisms were present, living and multiplying in the CNS or any other tissue or cavities.

D. Signs and Symptoms

The clinical symptoms are rather subtle and usually develop slowly. Some of the symptoms are irritability, confusion, hallucinations, dizziness, somnolence, drowsiness, and behavioral changes (Table 2). All of these symptoms may develop in patients with a background of chronic illness and debilitation without a history of recent swimming.[3,49,53] Usually, there are associated localizing signs such as seizures, hemiparesis, cranial nerve palsies, diplopia, aphasia, ataxia, and papilledema. The terminal event is deep coma. Some patients with GAE develop skin nodules a few days before the appearance of the neurological symptoms and signs.[47] Skin biopsies demonstrate chronic and granulomatous dermatitis with amebic trophozoites.

E. Differential Diagnosis

The isolation and identification of the trophozoite of *Acanthamoeba* spp. and their cysts from the CNS provides the only means of diagnosis. An early biopsy is stressed since the therapeutic success diminishes with time after onset of the disease. The examination of CSF may be of diagnostic value; however, the lumbar puncture may be contraindicated because of signs of increased intracranial pressure. Brain tissue and CSF should be cultured to isolate the amebic organisms. Frozen sections stained with H & E may result in a diagnosis within a few minutes after obtaining the specimen.

Other space occupying lesions of the CNS such as tumor (primary or metastatic), abscess, tuberculoma, or fungal infection must be considered in the differential diagnosis of GAE.

As previously mentioned, GAE is more likely to occur in the immunocompromised patient. It has been reported in patients iatrogenically immunosuppressed as well as in those who are debilitated from chronic disease.

The EEG tracings may be different from that of cerebral abscess. The neuroradiological findings provide a guide to the clinical neurologist and to the neurosurgeon as to the character of the lesion and its site and distribution. CT scans may show multiple

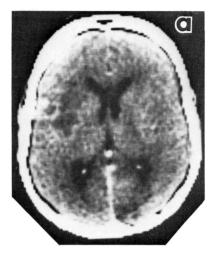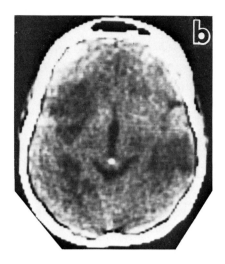

FIGURE 2.   (a) CT scan of the head with contrast in a case of GAE demonstrating multiple areas of decreased attenuation in the posterior frontal, anterior temporal, and right temporooccipital regions. (b) CT scan of the head with contrast of the same case of GAE showing large areas of decreased density in cerebral cortex and subcortical white matters of the frontal and parietooccipital lobes.

areas of decreased density in the cerebral cortex consistent with cerebral infarcts from septic emboli (Figures 2A and 2B). Brain biopsy may be required to establish the diagnosis and exclude other treatable lesions such as *Herpes simplex* meningoencephalitis.

When the symptoms and signs of both PAM and GAE are compared (Table 3), it is evident that localizing symptoms predominate in GAE.

Infection with *Acanthamoeba* spp. characteristically produces a well-developed granulomatous reaction with multinucleated giant cells. These features may be present to a lesser degree in some cases, probably as a consequence of long-term treatment with immunosuppressive drugs. It is possible that the host response to this protozoal infection may be influenced by prolonged immunosuppressive therapy.

When there is a deficiency in the production of macrophages, epitheloid cells, lymphocytes, plasma cells, giant cells, immunoblasts, fibroblasts, neutrophils, eosinophils, or granulocytes, the mechanism of granuloma formation is impaired. The ameba itself may induce macrophage toxicity which is an important factor in the production of the granulomatous reaction. When the function of the macrophages is inadequate and the irritant persists, granulomas cannot be formed. In addition, when there is a deficiency in cell-mediated immunity, a poor cellular response results and antibody production is impaired. The granulomatous reaction is an indicator of cell-mediated immunity. The essential factor in granuloma formation is a failure to eliminate the causative organism or irritant with subsequent persistence of the host defense mechanism. In some of the reported cases, the typical granulomatous reaction was not seen. In some cases with indisputable *Acanthamoeba* encephalitis, evidence of immunodeficiency or drug-induced immunosuppression cannot be established due to incomplete clinical and pathological documentation. It is difficult to make any conclusive statements from published reports regarding the underlying disease, the status of the cellular elements of immunity, or other factors that may have predisposed these patients to develop an opportunistic infection. Until more accurate data are accumulated, this aspect is left to speculation.

Table 3

PRESENTING SYMPTOMS/SIGNS IN 30 PATIENTS WITH AUTOPSY/
BIOPSY-PROVEN CNS INVOLVEMENT BY *ACANTHAMOEBA* SPP. (15
PATIENTS) AND *NAEGLERIA FOWLERI* (15 PATIENTS)

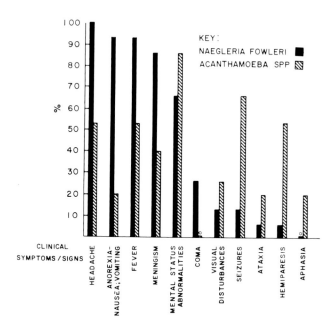

Table 4

GAE — PREDISPOSING FACTORS AND UNDERLYING DISEASES IN
OPPORTUNISTIC INFECTIONS

Predisposing factors	Basis for enhancing susceptibility or increased risk to infection
Allograft recipients	Impaired cell-mediated immunity
Chemotherapy	Impaired antibody synthesis
Bone marrow failure	Neutropenia, agranulocytosis
Splenectomy	Impaired IgM antibody production
Indwelling catheter or implanted devices	Foreign body
Radiotherapy, broad spectrum antibiotics, cortico-steroids, alcohol	Decreased antibody formation, immunosuppression, lymphoid and reticuloendothelial cell suppression

There are several predisposing factors that may be responsible for the development of opportunistic infections (Table 4). The clinical setting and underlying diseases associated with compromised host defense mechanisms are multiple. They include hospitalization, trauma, burns, surgery, metabolic diseases such as diabetes mellitus, liver disease, systemic lupus erythematosus, malignant tumors, blood transfusions, skin ulcers, pregnancy, lympho-, and hematoproliferative diseases. These conditions suppress or diminish the normal inflammatory and immune response of the host, both humoral and cell mediated.

The biological, morphological, clinical, and pathological differences between free-living, amphizoic, and parasitic amebas are stated in Table 5.

Acanthamoebic keratitis should be mentioned. This possibility should be suspected in patients who have had corneal ulcers which did not respond to the usual medica-

Table 5

FREE-LIVING OR AMPHIZOIC AND PARASITIC AMEBAS — COMPARISON AND CONTRAST

	Naegleria fowleri	*Acanthamoeba* sp.	*Entamoeba histolytica*
Protozoology	Trophozoites: 10—15 μm diam., round, conspicuous karyosome, clear, nuclear halo, culture grows 21—40°C, mitochondria; nuclear membrane remains intact during division; lobopodia; cyst: spherical	Trophozoites: round, dense; karyosome, clear nuclear halo; 25—35 μm diam, mitochondrias; nuclear membrane dissolved during division; acanthopodia; cyst: star shape with double walls	Trophozoitres: 10—50 μm diam; cytoplasm with erythrocytes; no mitochondria; pseudopodia, cyst: spherical with two or four nuclei 5—20 μm
Epidemiology	Good health — recent history of swimming in lake or swimming pool; hot summer months	History of poor health, immunological incompetent patient; no history of swimming	Oral ingestion of cysts in contaminated water or food; possible history of travel to warmer regions of world
Incubation (days)	3—7 days	Probably >10 days	Weeks
Portal of entry	Olfactory neuroepithelium	Skin, lung, olfactory neuroepithelium	Gastrointestinal tract
Onset	Fast	Slow; insidious	Slow
CNS spread	Direct; amyelinic nervous plexus	Probably hematogenous	Hematogenous from colon, lung, or liver
Organs affected	Brain, always	Brain, skin, eyes, lungs	Colon, liver, lung, brain
Clinical course	Acute, fulminant <10 days	Subacute (8—30 days); chronic (>32 days)	Chronic with acute episodes
Signs and symptoms	Headache; anorexia; nausea; vomiting; fever; meningism; mental abnormalities; diplopia; seizures	Mental abnormalities; seizures; fever; hemiparesis; headache; meningism; visual abnormalities	Depends on the location of lesions; diarrhea; abdominal pain; carriers: no symptoms; liver and pulmonary abscesses; CNS: increased intrapressure; headaches; papilledema.
Laboratory diagnosis and CSF	CSF similar to bacterial meningitis but sterile, neutrophilic pleocytosis, high protein; low glucose; direct exam of fresh CSF; trophozoites very active and motile; cultivation; inoculation in mice	Consistent with viral encephalitis and sterile cultivation; inoculation in mice	Serology, stool, and tissue exam; CSF: high protein; neutrophilic pleocytosis; no trophozoites or cysts
Host response	Purulent leptomeningitis, hemorrhagic necrotizing meningoencephalitis; brain edema; perivascular collection of amebas	Granulomatous encephalitis with focal necrosis and multinucleated giant cells; necrotizing angiitis	Colonic ulcers with acute inflammation; liver, lung, and brain abscesses
CNS amebic forms	Trophozoites	Trophozoites + cysts	Trophozoites + cysts
Differential diagnosis	Acute pyogenic (bacterial) leptomeningitis; viral encephalitis	Tuberculous, viral or fungal encephalitis; brain tumors; brain abscess	Ulcerative colitis; cholera; diverticulitis; acute appendicitis; brain tumors; abscesses
Therapy	Amphotericin B + Miconazole® + rifampin	Sulfadiazine? ketoconazole?	Metronidazole; tetracycline; diodohydroxyquinolin; chloroquine

Table 6
EXPERIMENTAL DRUGS USED IN FREE-LIVING AMEBIC INFECTIONS

Amphotericin B (Fungizone®) — effective against *N. fowleri* but nephotoxic in 89% of patients

Allopurinol® — effective in vitro against *Acanthamoeba* spp.

Clotrimazole, rifampicin, Baquaci® , and rifamycin (amebicidic in vitro against *Acanthamoeba* spp. Not effective against *N. fowleri*)

Hydroxystilbamidine — isethionate and 5-fluorocytosine (flucytosine) (antimetabolite) — amebicidal effect in vitro

Kanasulfin® — amebostatic and slight amebicidic against *Acanthamoeba* spp.

Sulfadiazine and gentamicin — inhibition of *Acanthamoeba* spp. in vitro

Tetracycline: inhibitory effect on the growth of *N. fowleri*; in vivo its effect is potentiated by amphotericin B

Paramycin (Humatin®), Miconazole® , and ketoconozole or R-41,400 — effective in vitro against strains of *Naegleria* and *Acanthamoeba* spp.

Polymyxin B® : inhibitory effect in vitro against *Acanthamoeba* spp.

Penicillin; rifamycin; cloroquine; Trimethoprim® ; nystatin; Carbarsone® ; chloramphenicol; clindamycin; neomycin; erythromycin; pentamidine; metronidazol; emetine; flagyl® ; Pyrimethamine; thiabendazole; nitrofurazone; concanavalin A; streptomycin; sulfamethoxazole — not effective against free-living amebas.

tions. The diagnosis of acanthamoebic keratitis should be suspected in patients who are posttraumatic. Corneal biopsy and corneal scrapings on culture usually permit the correct diagnosis.

F. Prognosis

The prognosis of GAE is grave, and in the majority of cases the diagnosis is made post-mortem. The causative organism has never been isolated from the CSF in living patients. E.C. Nelson isolated an *Acanthamoeba* spp. from the CSF of a boy who survived without neurological sequelae; however, the ameba was found 10 days after cultures were planted and could be considered a contaminant.

III. TREATMENT: PAM AND GAE: AN OVERVIEW (TABLE 6)

A. Primary Amebic Meningoencephalitis and Granulomatous Amebic Encephalitis

Emetine, penicillin G, metronidazole, chloroquine and other chemotherapeutic agents used in cases of *Enthamoeba histolytica* infection and bacterial meningitis are ineffective in free-living amebic infections.[12,13,15,28,32,61,82,83,87] Corticosteroids should not be given. To be effective, the treatment should be started early and associated with intensive, supportive care. However, until an antibiotic or drug with specific activity against free-living amebas with ability to reach therapeutic levels in CNS tissue is found, successful treatment is unlikely. Some of the drugs tested appear to be effective, but at higher concentrations which makes their use harmful to humans. Rifampin, clotrimazole, tetracycline, and Miconazole® have been shown to inhibit the growth of *Naegleria fowleri* in vitro.[39,59,75,103] Except for the customary supportive measures such as control of temperature and fluid and electrolyte management, specific therapy, in most cases, appears to have little influence on the natural course of this illness. Amphotericin B and Miconazole intravenously and intrathecally appear to be effective against *N. fowleri,* but only if given early in the course of the illness. A synergistic or additive effect of these drugs has been demonstrated in vitro.[94,98,100] Some patients have been reported to survive the infection with *N. fowleri* without sequelae.[95] The most reliable and effective treatment still appears to be intravenous and intrathecal amphotericin B and Miconazole in addition to rifampin; however, the toxicity of amphotericin B limits its practical value. Both amphotericin B and Miconazole have been shown effective in vitro against *Naegleria* species, while antimicrobials appear to be less active against *Acanthamoeba.*[36,37,43,66,104,108,110] Sulfadiazine has been shown to protect mice from infection by pathogenic strains of *Acanthamoeba* spp. *(Hartmannella)*[19,20,90] but has proved useless in human cases. Hydroxystilbamidine, 5-fluorocytosine, clotrimazole, tetracycline, ketoconazole, and gentamicin appear to be effective against *Acanthamoeba* spp.[11,14,17,29-31,33,60,106] *Acanthamoeba* spp. appears to be the most resistant strain of free-living amebas to treatment and can tolerate drugs very well.

One important factor regarding therapy is the ability of *Acanthamoeba* spp. to form cysts in tissue when the environment is unfavorable. The possibility of relapse occurs when the drugs reach the area where the ameba lives.[76] Evidently, further research to find suitable and effective chemotherapeutic agents or antibiotics is urgently needed to treat these grave diseases. Synergism has been demonstrated between tetracycline and amphotericin B in the mouse model of PAM.[104]

Inhibition of *N. fowleri* has been tested in vitro and in vivo. Growth inhibition in axenic culture was observed at concentrations of 100 μg/ml of pyrimethamine. Unfortunately, this drug cannot reach the CNS because it does not pass the blood-brain barrier even though they may reach high concentrations in the blood.

B. Acanthamoebic Keratitis

Ocular infections have been treated with ketoconazole (Nizoral®, Janssen Pharmaceutical, Inc.) both local and systemic, but this drug is still in the experimental stage. Amebic keratitis is very resistant to treatment, probably because of the ability of the ameba to encyst. There are no sufficient data on amebic ulcerations of the skin or even free-living amebic (acanthamoebic) pneumonitis. Pyrimethamine and corticosteroids in conjunction with triple sulfa pyrimidine and clindamycin have been used in *Toxoplasma gondii* infections, but there are no sufficient data on acanthamoebic keratitis. The correction of cellular or humoral deficiencies in those patients with immunologic incompetences may be logical, but the number of patients with *Acanthamoeba* spp. infection is so small that no conclusion can be reached at this time.

Some compounds like acriflavine, proflavine, hydroxystilbamidine isethianate, Miconazole®, (Ortho) paromomycin (Humatin® Parke Davis), amphotericin B (Fungizone®, Squibb), neomycin, and Polymyxin B® (Burroughts Wellcome) combined with Neosporin®, have demonstrated some amebicidal and amebostatic activity in ocular infections.[76] These compounds produce therapeutic effects by different mechanisms; for example, ketoconazole, which is an imidazole antifungal drug structurally related to Miconazole and Econazole, produces changes in the cell wall, increasing cell volume, abnormalities in cell division, and degeneration of subcellular organelles. Acriflavine (acridine) dye binds to mitochondrial DNA and inhibits the protein synthesis which is essential for cellular respiration. On the other hand, natamycin (pimaricin) is a polyene antibiotic that acts by disrupting the sterol configuration of the cell membrane by binding to ergosterol. This antibiotic produces holes in the membranes and allows the penetration of other drugs into the cytoplasm. Polymyxin B® disrupts the lipoproteins in the cellular membrane, enhancing the permeability. Imidazol is an antifungal agent that produces breaks in the cell membrane. Again, the majority of these compounds are still under investigation. More research is needed to prove their true therapeutic effects against acanthamoebic keratitis.

REFERENCES

1. Apley, J., Clark, S. K. R., Roome, A. P. Ch., Sandry, S. A., Saygi, G., Silk, B., and Warhurst, D. C., Primary amoebic meningoencephalitis in Britain, *Br. Med. J.*, 1, 596, 1970.
2. Blattner, R., Primary amoebic meningoencephalitis: infection with *Hartmannella (Acanthamoeba)*, *J. Pediatr.*, 70, 298, 1967.
3. Bhagwandeen, S. B., Carter, R. F., Naik, K. G., and Levitt, D., A case of hartmannellid amebic meningoencephalitis in Zambia, *Am. J. Clin. Pathol.*, 63, 483, 1975.
4. Bhatia, P. S., Roy, S., and Ahuja, G. K., Meningoencephalitis due to soil amoeba, *Neurology (India)*, 27, 44, 1979.
5. Brass, K., Meningoencefalitis amebiásica primaria (por *Naeglerias*) (Spanish), *Arch. Venez. Med. Trop. Parasitol. Med.*, 5, 291, 1973.
6. Butt, C. G., Primary amebic meningoencephalitis, *N. Engl. J. Med.*, 274, 1473, 1966.
7. Cain, A. R., Wiley, P. F., Brownell, D. B., and Warhurst, D. C., A fatal case of primary amoebic meningoencephalitis, *Arch. Dis. Childhood*, 56, 140, 1981.
8. Callicott, J. H., Jr., Amebic meningoencephalitis due to free-living amebas of the *Hartmannella (Acanthamoeba) — Naegleria* group, *Am. J. Clin. Pathol.*, 49, 84, 1968.
9. Carter, R. F., Sensitivity to amphotericin B of a *Naegleria* sp. isolated from a case of primary amoebic meningoencephalitis, *J. Clin. Pathol.*, 22, 470, 1969.
10. Carter, R. F., Primary amoebic meningoencephalitis: an appraisal of present knowledge, *Trans. R. Soc. Trop. Med. Hyg.*, 66, 193, 1972.
11. Casemore, D. P., Sensitivity of *Hartmannella-(Acanthamoeba)* to 5-fluorocytosine, hydroxystilbamidine, and other substances, *J. Clin. Pathol.*, 23, 649, 1970.
12. Červa, L., The effect of some drugs on the growth of the pathogenic strain of *Hartmannella (Acanthamoeba) castellanii* in vitro. A short communication, *Folia Parasitol. (Praha)*, 16, 357, 1969.
13. Červa, L., Sensitivity of pathogenic *Naegleria* to chemotherapeutics, *J. Protozool.*, 20 (Suppl.), 535, 1973.
14. Červa, L., *Naegleria fowleri:* trimethoprim sensitivity, *Science*, 209, 1541, 1980.
15. Chang, S., Resistance of pathogenic *Naegleria* to some common physical and chemical agents, *Appl. Environ. Microbiol.*, 35, 368, 1978.
16. Chang, S.L., Pathogenesis of pathogenic *Naegleria* amoeba, *Folia Parasitol. (Praha)*, 26, 195, 1979.
17. Childs, G. E. Diaminobenzidine reactivity of peroxisomes and mitochondria in a parasitic amoeba, *H. culbertsoni*, *J. Histochem. Cytochem.* 21, 26, 1973.
18. Cox, E. C., Amebic meningoencephalitis caused by *Acanthamoeba* species in a four month old child, *J. S. C. Med. Assoc.*, 76 (10), 459, 1980.
19. Culbertson, C. G., Overton, W. M., and Ensminger, P. W., Pathogenic *Hartmannella (Acanthamoeba)* and *Naegleria:* studies on experimental chemotherapy and pathology, in *Advances in Antimicrobial and Antineoplastic Chemotherapy*, Hejzlar, M., Semonský, M., and Masák, S., Eds., Urban & Schwarzenberg, Munich, 1972, 433.
20. Culbertson, C. G., Amebic meningoencephalitis, in *Antibiotics and Chemotherapy*, Vol. 30, Antiparasitic Chemotherapy, Schonfeld, H., Ed., S. Karger, Basel, 1981, 28.
21. Cursons, R. T. M., Brown, T. J., and Keys, E. A., Immunity to pathogenic free-living amoebae, *Lancet*, ii, 875, 1977.
22. Cursons, R. T. M., Keyes, E. A., Brown, T. J., Learmonth, J., Campbell, C., and Metcalf, P., IgA and primary amoebic meningoencephalitis, *Lancet*, 1, 223, 1979.
23. Cursons, R. T. M., Brown, T. J., Keys, E. A., Moriarty, K. M., and Till, D., Immunity to pathogenic free-living amoebae: role of humoral antibody, *Infect. Immun.*, 29, 401, 1980.
24. Cursons, R. T. M., Brown, T. J., Keys, E. A., Moriarty, K. M., and Till, D., Immunity to pathogenic free-living amoeba: role of cell mediated immunity, *Infect. Immun.*, 29, 408, 1980.
25. Darby, C. L., Conradi, S. E., Holbrook, T. W., and Chatellier, C., Primary amebic meningoencephalitis, *Am. J. Dis. Child.*, 133, 1025, 1979.
26. Das Gupta, A., Primary amoebic meningoencephalitis, *J. Ind. Med. Assoc.*, 54, 429, 1970.
27. Das, S. R. and Singh, B. N., Disease potential of free-living amoebae: virulence and chemotherapy of free-living amoebae, *J. Parasitol.*, 56, 67, 1970.
28. Das, S. R., Chemotherapy of experimental amoebic meningoencephalitis in mice infected with *Naegleria aerobia*, *Trans. R. Soc. Trop. Med. Hyg.*, 65, 106, 1971.
29. Dawson, M. W., Brown, T. J., and Till, D. G., The effect of Baquacil on pathogenic free-living amoebae (PFLA). I. In anexic conditions, *N.Z. J. Mar. Freshwater Res.*, 17, 305, 1983.
30. Dawson, M. W., Brown, T. J., Biddick, C. J., and Till, D. G., The effect of Baquacil on pathogenic free-living amoebae (PFLA). II. In simulated natural conditions; in the presence of bacteria and/or organic matter, *N.Z. J. Mar. Freshwater Res.*, 17, 313, 1983.
31. Dawson, M. W., Brown, T. J., Dibbick, C. J., and Till, D. G., The effect of Baquacil on pathogenic free-living amoebae (PFLA). III. Increased Baquacil concentration and exposure time in the presence of bacteria, *N.Z. J. Mar. Freshwater Res.*, in press.

32. DeCarneri, I. Sensibilita ai farmaci di amebe del suolo dei generei *Hartmannella e Naegleria,* Agenti etiologici di meningoencefaliti, *Rev. Parazitol.,* 31, 1, 1970.

33. Donald, J. J., Keys, E. A., Cursons, R. T. M., and Brown, T. J., Chemotherapy of primary amoebic meningoencephalitis (PAM), *N.Z. J. Med. Lab. Technol.,* 33, 23, 1979.

34. Dos Santos, J. G. Neto, M. D., Clinical laboratory and therapeutic aspects of primary amoebic meningoencephalitis, *G. Malattie Infettive Parassitarie,* 29, 706, 1977.

35. Duma, R. J., Ferrell, W. H., Nelson, E. C., and Jones, M. M. Primary amebic meningoencephalitis, *N. Engl. J. Med.,* 24, 1315, 1969.

36. Duma, R. J., In vitro susceptibility of pathogenic *Naegleria gruberi* to amphotericin B, *Antimicrob. Agents Chemother.,* 10, 109, 1970.

37. Duma, R. J. and Finley, R., In vitro susceptibility of pathogenic *Naegleria* and *Acanthamoeba* species to a variety of therapeutic agents, *Antimicrob. Agents Chemother.,* 10, 370, 1976.

38. Duma, R. J., Helwig, W. B., and Martinez, A. J., Meningoencephalitis and brain abscess due to a free-living amoebae, *Ann. Intern. Med.,* 88, 468, 1978.

39. Elmsly, C. J., Donald, J. J., Brown, T. J., and Keys, E. A., Chemotherapy of primary amoebic meningoencephalitis (PAM). II. Miconazole and R41,400 (Ketoconazole), *N.Z. J. Med. Lab. Technol.,* p. 37, 1980.

40. Feldman, M. R., *Naegleria fowleri:* fine structural localization of acid phosphatase and heme proteins, *Exp. Parasitol.,* 41, 290, 1977.

41. Ferrante, A. and Thong, Y. H., Antibody induced capping and endocytosis of surface antigens in *Naegleria fowleri, Int. J. Parasitol.,* 9, 599, 1979.

42. Ferrante, A. and Thong, Y. H., Unique phagocytic process in neutrophil mediated killing of *Naegleria fowleri, Immunol. Lett.,* 2, 37, 1980.

43. Ferrante, A., Comparative sensitivity of *Naegleria fowleri* to amphotericin B and amphotericin B methyl ester, *Trans. R. Soc. Trop. Med. Hyg.,* 76, 476, 1982.

44. Ferrante, A. and Mocatta, T. J., Human neutrophils require activation by mononuclear leukocyte conditioned medium to kill the pathogenic free-living amoeba, *Naegleria fowleri; Clin. Exp. Immunol.,* 56, 559, 1984.

45. García-Tamayo, J., González, J. E., and Martinez, A. J., Meningoencefalitis amibiana primaria y encefalitis granulomatosa amibiana, *Acta Med. Venez.,* 27, 84, 1980.

46. García-Tigera, J., Sotolongo-Guerra, F., Cepero-Noriega, F., Ibarra-Sanchez, E., and García-Ortega, J., Meningoencefalitis amebiana primaria. Estudio de un caso sospechoso y revisión de la literatura medica, *Rev. Cubana Med. Trop. (Havana),* 30, 161, 1978.

47. Grundy, R. and Blowers, R. A case of primary amoebic meningoencephalitis treated with Chloroquine, *E. Afr. Med. J.,* 47, 153, 1970.

48. Grunnet, M. L., Cannon, G. H., and Kushner, J. P., Fulminant amebic meningoencephalitis due to *Acanthamoeba, Neurology (N.Y.),* 31, 174, 1981.

49. Gullett, J., Mills, J., Hadley, K., Podemski, B., Pitts, L., and Gelber, R., Disseminated granulomatous *Acanthamoeba* infection presenting as an unusual skin lesion, *Am. J. Med.,* 67, 891, 1979.

50. Hecht, R. H., Cohen, A., Stoner, J., and Irwin, C., Primary amebic meningoencephalitis in California, *Calif. Med.,* 117, 69, 1972.

51. Hermanne, J., Jadin, J. B., and Martin, J. J., Meningo-encephalite amibienne primitive en Belgique. A propos de deux cas, *Ann. Pediatr.,* 19, 425, 1972.

52. Hermanne, J., Jadin, J. B., and Martin, J. J., Meningo-encephalite amibienne primitive en Belgique. Quatre premiers cas, *Acta Paediatr. Belg.,* 27, 348, 1973.

53. Hoffmann, E. O., Garcia, C., Lunseth, J., McGarry, P., and Coover, J., A case of primary amebic meningoencephalitis. Light and electron microscopy and immunohistologic studies, *Am. J. Trop. Med. Hyg.,* 27, 29, 1978.

54. Jadin, J. B., Hermanne, J., Robyn, G., Willaert, E., Van Maercke, Y., and Stevens, W., Trois cas de meningo-encephalite amibienne primitive observes a Anvers (Belgique), *Ann. Soc. Belge Med. Trop.,* 51, 255, 1971.

55. Jadin, J. B. and Willaert, E., Trois cas de meningo-encephalite amibienne a *N. gruberi* observes a Anvers (Belgique), *Protistologica,* 8, 95, 1972.

56. Jadin, J. B., Hermanne, J., and Willaert, E., La meningo-encephalite ambienne primitive, *Med. Maladies Infect.,* 2, 205, 1972.

57. Jadin, P. J. B., Meningoencephalitis from free-living amoeba of *Naegleria-Acanthamoeba* genus. Introduction, *G. Malatti Infettive Parassitarie,* 29, 643, 1977.

58. Jakovljevich, R. and Talis, B., Recovery of a hartmanneloid ameba in the purulent discharge from a human ear (abstr.) *J. Protozool.,* 16 (Suppl), 1969.

59. Jamieson, A., Effect of clotriamazole in *Naegleria fowleri, J. Clin. Pathol.,* 28, 446, 1975.

61. Krainik, F., Merle, G., and Bertin, M., Will *Naegleria fowleri* become a public Health Problem? Sem. Hop. (Paris) 59, 775, 1983.

62. Lam, A. H., deSilva, M., Procopis, P., and Kan, A., Primary amoebic *(Naegleria)* meningoencephalitis. Case report, *J. Comput. Assist. Tomogr.*, 6, 620, 1982.

63. Lawande, R. V., John, I., Dobbs, R. H., and Egler, L. J., A case of primary amebic meningoencephalitis in Zaria, Nigeria, *Am. J. Clin. Pathol.*, 71, 591.

64. Lawande, R. V., Duggan, M. D., Constantinidou, M., and Tubbs, D. B., Primary amoebic meningoencephalitis in Nigeria (report of two cases in children), *J. Trop. Med. Hyg.*, 82, 84, 1979.

65. Lawande, R. V., MacFarlane, J. T., Weir, W. R. C., and Awunor-Renner, C., A case of primary amebic meningoencephalitis in a Nigerian farmer, *Am. J. Trop. Med. Hyg.*, 29, 21, 1980.

66. Lee, K. K., Karr, S. L., Wong, M. M., and Hoeprich, P. D., In vitro susceptibilities of *Naegleria fowleri* strain HB-1 to selected antimicrobial agents, singly and in combination, *Antimicrob. Agents and Chemother.*, 16, 217, 1979.

67. Malhotra, K. K., Kakar, P. N., Pillay, P., Pathak, L. R., and Chuttani, H. K., Unusual presentation of primary amoebic meningoencephalitis — a serious diagnostic and therapeutic problem, *J. Trop. Med. Hyg.*, 81, 113, 1978.

68. Markowitz, S.M ., Martinez, A. J., Duma, R. J., and Shiel F'O. M., Myocarditis associated with primary amebic *(Naegleria)* meningoencephalitis, *Am. J. Clin. Pathol.*, 62, 619, 1974.

69. Martinez, A. J., Amebic meningoencephalitis due to *Naegleria* and *Acanthamoeba*. Cliniconeuropathological correlation, in *Proc. Int. Conf. on Amebiasis*, Sepulveda, B. and Diamond, L. S., Eds., Instituto Mexicano del Seguro-Social, Mexico City, 1975, 64.

70. Martinez, A. J., Dos Santos, J. G., and Nelson, E. C., *Naegleria* and *Acanthamoeba/Hartmannella* sp. The causative organism and the diseases produced: compared and contrasted, *Lab. Invest.*, 34, 39, 1976.

71. Martinez, A. J., Sotelo-Avila, C., Garcia-Tamayo, J., Takano Moron, J., Willaert, E., and Stamm, W. P., Meningoencephalitis due to *Acanthamoeba* sp. Pathogenesis and clinicopathological study, *Acta Neuropathol. (Berl.)*, 37, 183, 1977.

72. Martinez, A. J. and Janitschke, K., Amobenzephalitis durch *Naegleria* und *Acanthamoeba*. Vergleich und Gegenuberstellung der organismen und der Erkrankungen, *Immun. Infekt.*, 7, 57, 1979.

73. Martinez, A. J. and Amado-Ledo, D. E., Meningoencefalitis y encefalitis producidas por amebas de vida libre. Protozoologia, epidemiologia, y neuropatologia, *Morfol. Normal Patol. (Spanish) (Granada)*, 3, 679, 1979.

74. Martinez, A. J. and Kasprzak, W., Patogenne pelzaki wolnozyjaceprzeglad, *Wiadomosci Parazytol. (Polish)*, 26, 495, 1980.

75. Mehta, A. P. and Guirges, S. Y., Acute amoebic dysentery due to free-living amoebae treated with metronidazole, *J. Trop. Med. Hyg.*, 87, 134, 1979.

76. Nagington, J. and Richards, J. E., Chemotherapeutic compounds and *Acanthamoeba* from eye infections, *J. Clin. Pathol.*, 29, 648, 1976.

77. Nakamura, T., Kobayashi, M., Wada, H., Tsunoda, Y., Akai, K., and Omata, K., An autopsy case of primary amebic meningoencephalitis, *Adv. Neurol. Sci. (Japanese)*, 23, 500, 1979.

78. Nicoll, A. M., Fatal primary amebic meningoencephalitis, *N.Z. Med. J.*, 78, 108, 1973.

79. Pan, N. R. and Ghosh, T.N., Primary amoebic mengoencephalitis in two Indian children, *J. Ind. Med. Assoc.*, 56, 134, 1971.

80. Pernin, P., Cuillert, C., and Lious, C., Free-living amoebae pathogenic for man, *Lyon Med.*, 245 (Suppl. 10), 105, 1981.

81. Powers, J. S., Abbott, R., Boyle, L., et al., One case of PAM in California; one case of PAM in Florida; one case GAE in New York, *Morbidity Mortality Wkly. Rep.*, 27, 343, 1978.

82. Prasad K. B. N., In vitro effect of drugs against pathogenic and nonpathogenic free-living amoebae and on anaerobic amoebae, *Ind. J. Exp. Biol.* 10, 43, 1972.

83. Pringle, H. L., Bradley, S. G., and Harris, L. S., Susceptibility of *Naegleria fowleri* to Δ^9. Tetrahydrocannabinol, *Antimicrob. Agents Chemother.*, 6, 674, 1979.

84. Reyes, P. F., Malherbe, H. H., Bertoni, J. J., and Parker, J. C., Histopathologic and in vitro culture studies in a case of primary amebic meningoencephalitis (abstr.), *J. Neuropathol. Exp. Neurol.*, 41, 380, 1980.

85. Ringsted, J., Jager, B. V., Suk, D., and Visvesvara, G. S., Probable *Acanthamoeba* meningoencephalitis in a Korean child, *Am. J. Clin. Pathol.*, 66, 723, 1976.

86. Robert, V. B. and Rorke, L. B., Primary amebic encephalitis, probably from *Acanthamoeba, Ann. Intern. Med.*, 79, 174, 1973.

87. Rondanelli, E. G., Carosi, G., Minoli, L., and Filice, G., Le meningoencefaliti amebiche primarie (MAP) da amebe del grupo *Hartmannella-Naegleria*. Un capitol nuovo de patologia amebica, *Terapia*, 57, 136, 1972.

88. Rothrock, J. F. and Buchsbaum, H. W., Primary amebic meningoencephalitis, *J. A. M. A.*, 243, 2329, 1980.

89. Rowan-Kelly, B., Ferrante, A., and Thong, Y. H., Activation of complement by *Naegleria, Trans. R. Soc. Trop. Med. Hyg.*, 74, 333, 1980.

90. Rowan-Kelly, B., Ferrante, A., and Thong, Y. H., The chemotherapeutic value of sulphadiazine in treatment of *Acanthamoeba* meningoencephalitis in mice, *Trans. R. Soc. Trop. Med. Hyg.,* 76, 636, 1982.

91. Rowan-Kelly, B. and Ferrante, A., Immunization with killed *Acanthamoeba culbertsoni* antigen and amoeba culture supernatent antigen in experimental *Acanthamoeba* meningoencephalitis, *Trans. R. Soc. Trop. Med. Hyg.,* 78, 179, 1984.

92. Salles-Gomes, C. E., Barbosa, E. R., Nobrega, J. P., Scaff, M., and Spina-Franca, A., Primary amoebic meningoencephalomyelitis: a case report (Portuguese), *Arq. Neuropsiquiatria,* 36, 139, 1978.

93. Saygi, G., Warhurst, D. C., and Roome, A. P. C. H., A study of amoebae isolated from the Bristol cases of primary amoebic encephalitis, *Proc. R. Soc. Med.,* 66, 277, 1973.

94. Schuster, F. L. and Rechthand, E., In vitro effects of amphotericin B on growth and ultrastructure of the amoeboflagellates, *Naegleria gruberi* and *Naegleria fowleri, Antimicrob. Agents Chemother.,* 8, 591, 1975.

95. Seidel, J. S., Harmatz, P., Visvesvara, G. S., Cohen, A., Edwards, J., and Turner, J., Successful treatment of primary amebic meningoencephalitis, *N. Engl. J. Med.,* 306, 346, 1982.

96. Sixl-Voigt, B. and Sixl, H., Specific complement fixation reaction for *Acanthamoeba, Z. Immunitaetsforsc.,* 142, 248, 1971.

97. Sotelo-Avila, C., Taylor, F. J., and Ewing, C. W., Primary amebic meningoencephalitis in a healthy 7 year old boy, *J. Pediatr.,* 85, 131, 1974.

98. Stevens, A. R. and O'Dell, W. D., In vitro and vivo activity of 5-fluorocytosine on *Acanthamoeba, Antimicrob. Agents Chemother.,* 6, 282, 1974.

99. Stevens, A. R., Kilkpatrick, T., Willaert, E., and Capron, A., Serologic analyses of cell-surface antigens of *Acanthamoeba* spp. with plasma membrane antisera, *J. Protozool.,* 24, 316, 1977.

100. Stevens, A. R. and Willaert, E., Drug sensitivity and resistance of four *Acanthamoeba* species, *Trans. R. Soc. Trop. Med. Hyg.,* 74, 806, 1980.

101. Stevens, A. R., Shulman, S. T., Lansen, T. A., Cichon, M. J., and Willaert, E., Primary amebic meningoencephalitis: a report of two cases and antibiotic and immunologic studies, *J. Infect. Dis.,* 143, 193, 1981.

102. Symmers, W. St. C. Primary amoebic meningoencephalitis in Britain, *Br. Med. J.,* 4, 449, 1969.

103. Thong, Y. H., Rowan-Kelly, B., Shepherd, C., and Ferrante, A., Growth inhibition of *Naegleria fowleri* by tetracycline, rifamycin and miconazole, *Lancet,* ii, 876, 1977.

104. Thong, Y. H., Rowan-Kelly, B., Ferrante, A., and Shepherd, C. Synergism between tetracycline and amphotericin B in experimental amoebic meningoencephalitis, *Med. J. Aust.,* 1, 663, 1978.

105. Thong, Y. H., Shepherd, C., Ferrante, A., and Rowan-Kelly, B., Protective immunity to *Naegleria fowleri* in experimental amebic meningoencephalitis, *Am. J. Trop. Med. Hyg.,* 27, 238, 1978.

106. Thong, Y. H., Rowan-Kelly, B., and Ferrante, A., Pyrimethamine in experimental amoebic meningoencephalitis, *Aust. Paeditr. J.,* 14, 177, 1978.

107. Thong, Y. H., Ferrante, A., Rowan-Kelly, B., and O'Keefe, D. E., Immunization with culture supernantant in experimental amoebic meningoencephalitis, *Trans. R. Soc. Trop. Med. Hyg.,* 73, 684, 1979.

108. Thong, Y. H., Rowan-Kelly, B., and Ferrante, A., Treatment of experimental *Naegleria* meningoencephalitis with a combination of Amphotericin B and Rifamycin, *Scand. J. Infect. Dis.,* 11, 151, 1979.

109. Thong, Y. H., Ferrante, A., Rowan-Kelly, B., and O'Keefe, D., Immunization with live amoebae, amoebic lysate and culture supernatant in experimental *Naegleria* meningoencephalitis, *Trans. R. Soc. Trop. Med. Hyg.,* 74, 570, 1980.

110. Thong, Y. H. and Seidel, J., Chemotherapy for primary amebic meningoencephalitis (Letters to the Editor), *N. Engl. J. Med.,* 306, 1295, 1982.

111. Vonoczky, J., Amoben-meningoencephalitis: Heilung durch pyrimethamin, *Acta Paediatr. Acad. Sci. Hung.,* 19, 45, 1978.

112. Warhurst, D. C., Roome, A. P., and Saygi, G., *Naegleria* sp. from human cerebrospinal fluid, *Trans. R. Soc. Trop. Med. Hyg.,* 64, 19, 1970.

113. Wessel, H. B., Hubbard, J., Martinez, A. J., and Willaert, E., Granulomatous amebic encephalitis (GAE) with prolonged clinical course: CT scan findings, diagnosis by brain biopsy, and effect of treatment (Abstr.) *Neurology (Minneap.),* 30, 442, 1980.

114. Yamauchi, T., Jimenez, J. F., McKee, T. W., Euler, R., and White, P. C., Amebic meningoencephalitis in Arkansas, *J. Ark. Med. Soc.,* 76, 164, 1979.

115. Zimak, V. and Ferdinandova, M., Clinical picture of suppurative meningoencephalitides probably caused by an amoeba of the genus *Hartmannella* (in Czech.), *Casopis Lekaru Ceskych (Praha),* 107, 724, 1968.

Chapter 6

ANATOMO-PATHOLOGICAL CHARACTERISTICS: PAM AND GAE

I. ACUTE INFECTION DUE TO *NAEGLERIA* SPP: PAM (Figure 1)

A. Macroscopic Appearance of Infected CNS

In contrast to *Entamoeba histolytica* that may produce brain abscesses,[2,6,40,57] free-living amebas like *Naegleria fowleri* usually produce a more diffuse and fulminant meningoencephalitis.[7,11,38]

The cerebral hemispheres are usually soft, markedly swollen, edematous, and severely congested with evidence of increased intracranial pressure; uncal and cerebellar tonsillar herniations might be present (Figure 2). They may weigh more than 1500 g.[13,29,30]

The leptomeninges (arachnoid and pia mater) are diffusely hyperemic and opaque with scant purulent exudate within some sulci, base of the brain, interpeduncular fossa, basal cisterna, Sylvian fissure, brainstem, and cerebellum. The cranial nerves are usually encompassed by a modest amount of sanguino-purulent exudate. The pituitary gland usually reveals areas of recent necrosis.

The olfactory bulbs are markedly hemorrhagic, necrotic, and friable, surrounded by a modest quantity of purulent exudate (Figure 3).[25] As the pathogenic amebas reach the subarachnoid space and the brain through the olfactory neuroepithelium and the fascicles of the olfactory nerves, amebic rhinitis is an associated finding (Figure 4). Amebic trophozoites may be found within fascicles of the unmyelinated axons of the olfactory nerves and within the nasal mucosa (Figures 5 and 6). Modest acute inflammatory exudate is an associated finding in and around the olfactory nerves. The cribriform plate, the piercing unmyelinated nerve fibers, and the dura mater of the supraorbital area should be examined for inflammatory reaction and for amebic trophozoites.

The cortex shows numerous superficial hemorrhages which sometimes penetrate deeply in subcortical white matter. The cortex bares the brunt of the attack (Figures 7A, B, C, and D). The cortical involvement varies from region to region in regard to the intensity and extent of the lesions. Most of the lesions are in and around the base of the orbitofrontal and temporal lobes, base of the brain, basal cisternas, hypothalamus, midbrain, pons, medulla oblongata, cerebellar folia (Figure 8), and upper portion of the spinal cord. The parietal and occipital regions might also be affected but in lesser degree. Sometimes areas of necrosis and hemorrhage may be found in the cingulate gyri or even in the ventricular walls or within the periaqueductal gray matter.

B. Microscopic Findings in Infected CNS

The cortical gray matter, mainly at the base of the brain, is the site of predilection with severe involvement in all cases. Fibrinous-purulent, leptomeningeal exudate containing numerous polymorphonuclear leukocytes, few eosinophils, scant macrophages, and some lymphocytes may be present throughout cerebral hemispheres, brainstem, cerebellum, and upper portions of spinal cord (Figure 9).[12]

Focal, recent thrombosis, necrotizing angiitis, and fibrinoid necrosis are occasionally seen. Amebas may be seen in scant number in the purulent exudate of the subarachnoid space (Figures 10, 11, 12), some of them being phagocytosed by polymorphonuclear leukocytes.[21,22] Pockets of numerous trophozoites may be seen within edematous and necrotic neural tissue, with scant or no acute inflammatory cells. Foci of demyelination may be present in areas far away from the most affected ones.[22] The

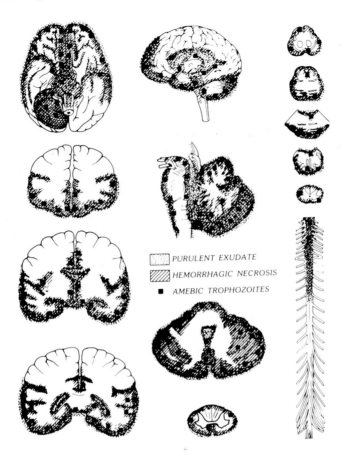

PURULENT EXUDATE

HEMORRHAGIC NECROSIS

■ AMEBIC TROPHOZOITES

FIGURE 1.   Primary amebic meningoencephalitis (PAM) due to *Naegleria* spp. Gross and microscopic findings in cerebrum, brainstem, cerebellum, and spinal cord. (From Martinez, A. J., *Neurology*, 30, 567, 1980. With permission.)

trophozoites are usually located in the adventitia and perivascular spaces[61] of small- and middle-size arteries and arterioles. Venules and capillaries may be surrounded by amebic trophozoites. No cysts are present within CNS lesions (Figure 10).[14,20]

Through the foramina of Luschka and Magendie, the amebas can enter the ventricular system, reach the choroid plexuses, and destroy the ependymal lining of the third, fourth, and lateral ventricles.[43] Choroid plexitis may be an associated finding.

Deep in Virchow-Robin spaces well-preserved amebic trophozoites may be found in large number, chiefly grouped around blood vessels and apparently provoking minimal or no inflammatory response (Figure 11). The invading amebas follow a centripetal invasion, from subarachnoid spaces and along Virchow-Robin spaces to deep gray and white matter. Inflammatory cells and occasional macrophages showing varying degrees of degeneration may be found. Phagocytosed CNS material, erythrocytes, and myelin fragments may be seen within amebas.[44] Extensive hemorrhagic necrosis of neural tissue involving neurons, astrocytes, and oligodendrocytes may be seen without any inflammatory cellular reaction. In better preserved areas microglia proliferation may be present.

It has been reported that cases dying of PAM may show associated diffuse or focal myocarditis. The pathogenesis of this myocarditis is unclear, but, perhaps, autonomic mechanisms (sympathetic and parasympathetic) might be involved in its pathogenesis. No trophozoites or cysts were found in the myocardial lesions.[43]

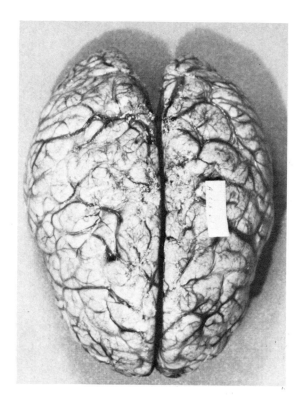

FIGURE 2.   PAM due to *Naegleria fowleri*. The cerebral hemispheres are swollen, congested with scant purulent exudate within sulci. (From Medical College of Virginia; A-338-67. With permission.)

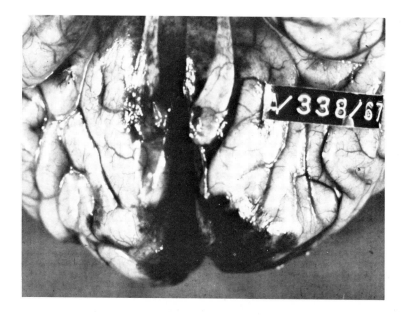

FIGURE 3.   Base of the brain, showing focal necrosis and edema of the olfactory bulbs and hemorrhagic foci in the orbitofrontal area. There is modest sanguino-purulent exudate within sulci around congested blood vessels. (From Medical College of Virginia; A-338-67. With permission.)

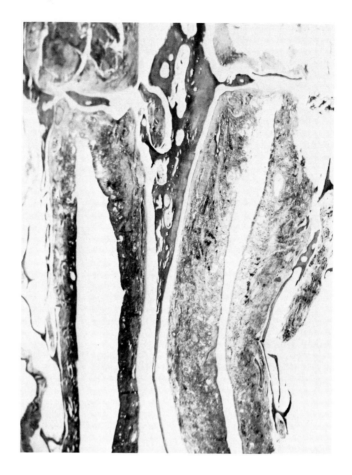

FIGURE 4.   Coronal section passing through the cribriform plate, the nasal septum, and the nasal cavities. The right olfactory mucosa and neuroepithelium is markedly swollen due to *Naegleria fowleri* infection. The portal of entry by which the amebas may reach the subarachnoid space and the central nervous system is by this route (27-year-old white man that swam 4 days before becoming sick). (H & E; magnification × 10.) (From Medical College of Virginia, A-338-67. With permission.)

C. Differential Diagnosis

Free-living amebas, as etiological agents, should be suspected and included in the list of differential diagnosis in every case of purulent, pyogenic (suppurative) leptomeningitis in which bacteria are not found. The histological diagnosis of PAM usually is not difficult provided that the the pathologists are aware of this possibility. CSF should be sent to the microbiology laboratory for culture, isolation of amebas, and inoculation into mice, particularly if amebic trophozoites have been found by direct microscopic examination or after staining the CSF smear with Giemsa or Wright stain techniques. Gram stain smear of the CSF sediment is of no value. A drop of the CSF using light microscopy with decreased light (lowering the condenser or closing the light apperture) or using phase contrast equipment may reveal motile amebas. However, the amebic trophozoite may be confused with macrophages or other motile mononuclear cells from the blood. In tissue sections amebic trophozoites may mimic macrophages and they should be differentiated from mononuclear and other host cells.

The morphological characteristics of *N. fowleri*, *N. australiensis*, *N. jadini*, and *N. lovaniensis* trophozoites include a centrally placed, spherical, dense nucleolus or kar-

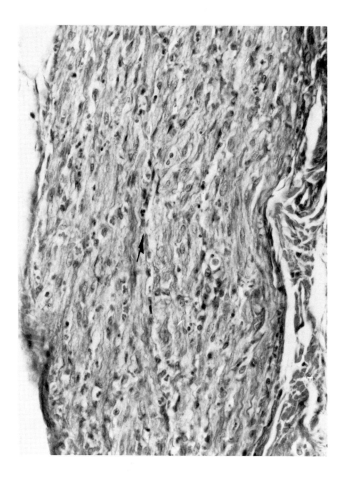

FIGURE 5.   Submucosal nervous plexus, from the same case as Figure 4, containing one amebic trophozoite (arrow) and modest lymphocytic infiltrate. (H & E; magnification × 100.)

yosome surrounded by a clear nuclear halo and abundant granular and vacuolated cytoplasm. The size of the trophozoites is between 12 to 18 μm.

II. CHRONIC "GRANULOMATOUS" INFECTION DUE TO *ACANTHAMOEBA* SPP. GAE (FIGURE 13)

A. Macroscopic Appearance of Infected CNS

The cerebral hemispheres, cerebellum, and brainstem are characterized by focal and isolated hemorrhagic softening associated with edema slightly overlying the surface of the convexity.[32,35] There is flattening of the gyri and narrowing of sulci in the affected areas (Figures 14, 15, and 16). Uncal notching and cerebellar tonsillar herniations may be a prominent feature. The brain might weight more than 1300 g. The leptomeninges, mainly over the affected cortical areas, are congested, opaque, and adhere to the cerebral cortex.[17,18,45] The rest of the leptomeninges are transparent and with normal vascularity (Figures 14 to 17). The diencephalon, deep thalami, other basal ganglia, and brainstem are usually the most affected areas [26,27,31] and showing foci of hemorrhagic necrosis (Figures 17A, B, C, and D and 18). The leptomeninges and cerebral cortex at the base of the cerebral hemispheres are usually clean and not affected. The blood vessels of the circle of Willis are intact and not involved in the inflammatory process. The cranial nerves are usually not involved in the inflammatory process.[19] The pituitary

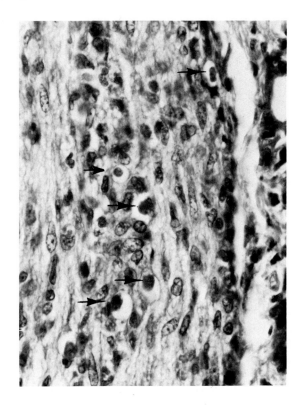

FIGURE 6.   Amebic trophozoites (arrows) within unmyelinated submucosal olfactory nerve bundles. Same case as Figure 5. (H & E; magnification × 500.)

gland is, likewise, not involved. The olfactory bulbs and spinal cord are usually spared and without any damage.

B. Microscopic Findings in Infected CNS

The leptomeninges are focally and minimally inflamed   always over the lesions of the cerebral cortex, cerebellar, or brainstem.

The modest chronic inflammatory exudate covering the cortical gray matter is composed of mostly lymphocytes, monocytes, few plasma cells, histiocytic elements, and occasionally few polymorphonuclear leukocytes.[15,23,46,48]

Usually the CNS lesions of GAE are characterized by necrosis with foci of hemorrhages and localized leptomeningitis (Figure 19). Amebic trophozoites and cysts are usually found within the most affected areas and around blood vessels, and piercing their walls (Figures 20 and 21). The gray and white matter may reveal a granulomatous reaction, with multinucleated giant cells[45] and astrocytosis (Figure 22). Some cases under immunosuppression and with impaired immune mechanisms may not have multinucleated giant cells. Angiitis with necrotizing arteritis and fibrinoid necrosis are frequently seen. The invading amebas follow a centrifugal invasion, usually from deep gray matter nuclei or central areas of the brainstem to the brain surface or cerebral cortex.[45,49,50,54]

Trophozoites and cysts are the amebic forms observed, scattered throughout the lesions, but preferentially located in perivascular spaces and invading blood vessel walls (Figures 20 and 21). Both amebic trophozoites and cysts with the typical star-shape or spherical with double walls may be found piercing the walls of small- and middle-size arteries from the adventitia to the muscularis (Figure 21). As the blood

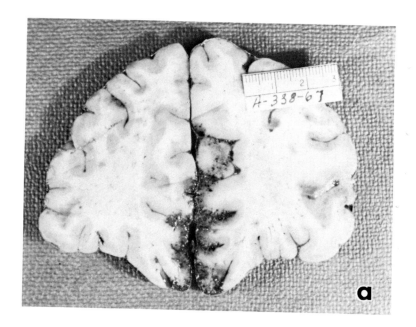

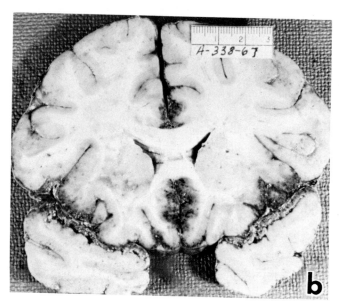

FIGURE 7. (a, b, c, and d) Primary amebic meningoencephalitis due to *Naegleria fowleri*. Coronal sections of the cerebral hemispheres showing the topographic distribution of the lesions. (From Medical College of Virginia; A-338-67. With permission.)

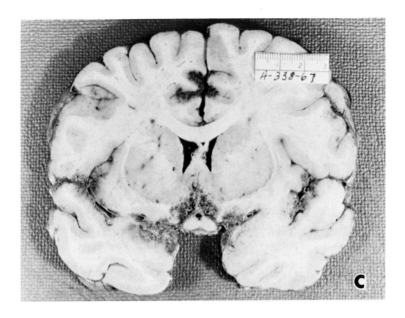

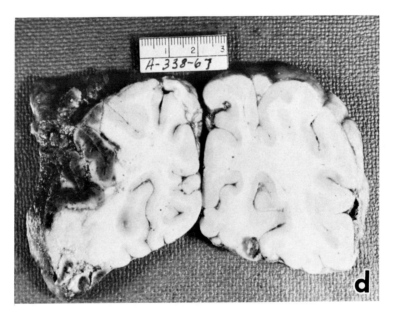

FIGURE 7 (continued)

vessel wall may weaken, "mycotic" aneurysmal formation and thrombosis may be present.[48-50]

The presence of amebic cysts in GAE denotes that even debilitated human hosts reveal a fighting response against the protozoan. The encystation phenomenon is not seen in PAM.

C. Microscopic Findings in Affected Tissues Other Than CNS

Chronic ulcerations of the skin due to *Acanthamoeba* spp. have been reported (Figure 23).[3,27,36] In fact, this appears to be the portal of entry of the protozoa into the body. Apparently, the amebic trophozoites may reach the CNS by hematogenous spread. The skin and other mucosal or pulmonary lesions are characterized by a subacute and chronic "granulomatous" inflammation in which multinucleated giant cells,

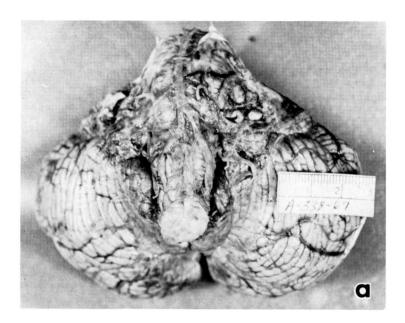

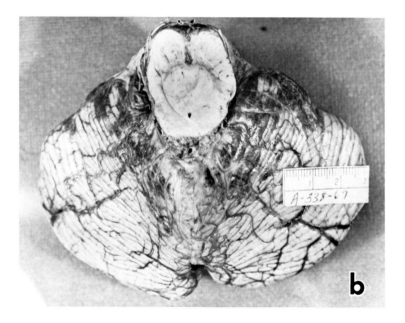

FIGURE 8. (a and b) Primary amebic meningoencephalitis. External appearance of the brain-stem and cerebellum. There is opacity of leptomeninges in the cerebellopontine angles, upper portion of the vermis, and the cisterna magna. (c) Sagittal section through the cerebellar vermis and (d) parasagittal section through the cerebellar hemisphere. There are multiple areas of hemorrhagic necrosis. (From Medical College of Virginia; A-338-67. With permission.)

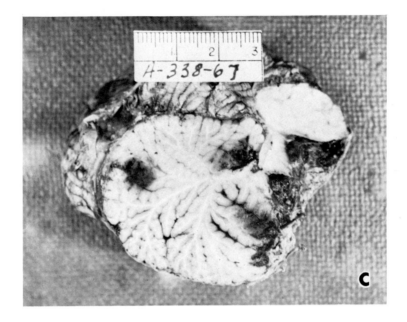

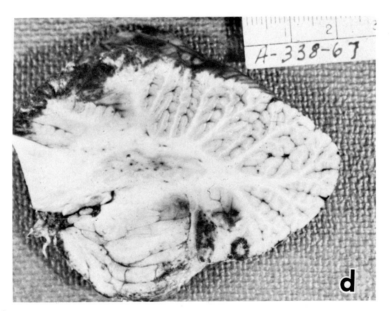

FIGURE 8 (Continued)

trophozoites, and cysts may be present. In some cases of GAE lesions with "granulomas" and multinucleated foreign body, giant cell has been found in skin (Figure 24), prostate, kidneys, uterus, and pancreas.[45,48] These lesions were probably the result of hematogenous dissemination from a primary focus, probably in the skin, lungs, or maybe from the "secondary" CNS lesions.[50] "Amebic pneumonitis" is characterized by areas of consolidation of pulmonary parenchyma (Figure 25). Microscopically amebic trophozoites and cysts (Figure 26) may be found frequently associated with other "opportunist" pathogens: bacteria, fungi, or viruses. Abscesses in oral tissues containing amebic trophozoites and cysts have been reported.[4,24,39,53] Ear infections have also been reported.[39] Ultrastructural studies may be done using formalin-fixed tissue and embedded in plastic. The fine details of cysts and trophozoites may be seen (Figure 27).

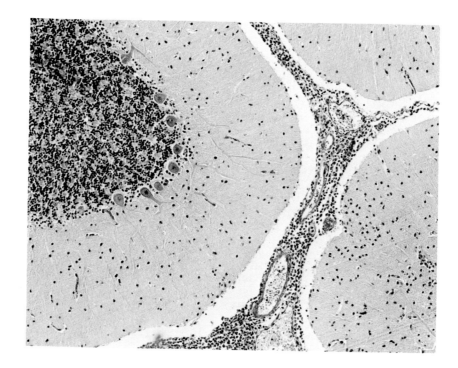

FIGURE 9. Acute inflammatory exudate within cerebellar folia. There were few small foci with inflammatory cells in the molecular layer. (H & E; magnification × 100.) (From Medical College of Virginia; A-338-67. With permission.)

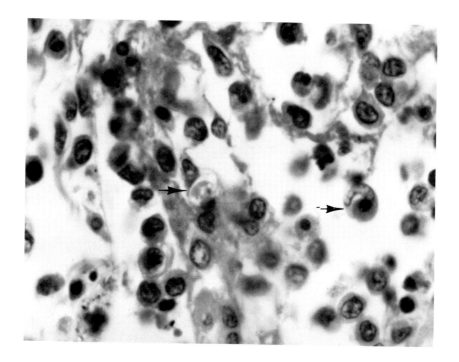

FIGURE 10. Amebic trophozoites within inflammatory cells. Some trophozoites have been phagocytosed by macrophages (arrows). (H & E; magnification × 1000.) (From Medical College of Virginia; A-338-67. With permission.)

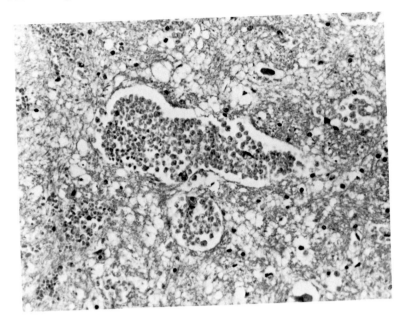

FIGURE 11.   Primary amebic meningoencephalitis due to *Naegleria fowleri*. Clusters of amebic trophozoites filling perivascular spaces. There is conspicuous absence of inflammatory exudate. (H & E; magnification × 250.) (Case of PAM courtesy of Rosa Fiol, M. D., Neuropathologist, University of Puerto Rico.)

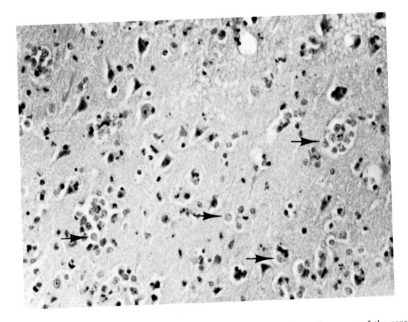

FIGURE 12.   Group of trophozoites of *Naegleria fowleri* in a perivascular space of the cerebral cortex. The trophozoites have a single clear nucleus with a centrally placed dense karyosome. (Arrows) (H & E; magnification × 500.) (From Medical College of Virginia; A-338-67. With permission.)

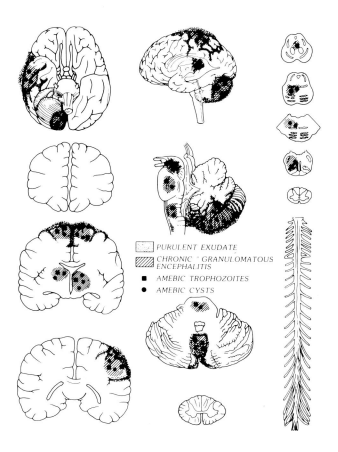

PURULENT EXUDATE

CHRONIC GRANULOMATOUS ENCEPHALITIS

AMEBIC TROPHOZOITES

AMEBIC CYSTS

FIGURE 13. Granulomatous amebic encephalitis (GAE) due to *Acanthamoeba* spp. Gross and microscopic findings in cerebrum, brainstem, cerebellum, and spinal cord. (From Martinez, A.J., *Neurology,* 30, 567, 1980. With permission.)

D. Differential Diagnosis

Acanthamoeba spp. must be considered in the differential diagnosis of any chronic suppurative dermatological infection, especially in an immunocompromised hosts. The histological appearance of acanthamoebic inflammation may mimic deeper fungal infection, chronic bacterial infection, acid fast bacilli infection, or a foreign body inflammation.

Acanthamoeba spp. should be differentiated from *Naegleria* spp. Usually the lesions due to *Acanthamoeba* spp. are accompanied by spherical or stellate cysts with double wall. The trophozoites of *Acanthamoeba* spp. are usually larger (15 to 45 μm) than *Naegleria* spp. In addition, the trophozoites of *Acanthamoeba* spp. must be differentiated from host cells, especially macrophages. The vegetative or trophic forms of *Entamoeba histolytica* are usually easy to separate from free-living amebas. They are usually larger (about 70 μm in diameter) with a nucleus containing prominent peripheral nuclear chromatin and delicate granules of chromatin between the periphery of the nucleus and the central karyosome. The cysts usually possess four nuclei. Brain tissue and CSF should be cultured for the isolation and identification of the protozoa.[16]

Immunofluorescent and immunoperoxidase techniques may be used to precisely identify the type of ameba responsible for the encephalitic process, particularly when the diagnosis was not made when the patient was still alive.

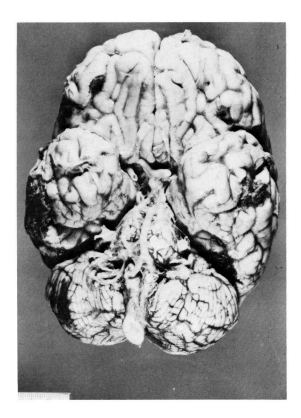

FIGURE 14.   Granulomatous amebic encephalitis (GAE) due to *Acanthamoeba castellanii.* Basal view of the cerebral hemispheres, brainstem, cerebellum, and upper portion of the spinal cord. The orbitofrontal area and olfactory bulbs and nerves are clean. Foci of hemorrhagic encephalomalacia are noted. (From Presbyterian-University Hospital of Pittsburgh; PA-80-28. With permission.)

III. OPHTHALMIC INFECTIONS DUE TO *ACANTHAMOEBA* SPP.

Ocular infection with *Acanthamoeba* spp. is presumably acquired from direct invasion of the ocular tissues. Several cases have been reported of chronic corneal ulceration due to *Acanthamoeba* spp. that failed to respond to the usual antibacterial, antifungal, or antiviral treatment.[1,5,28,33,37,41,51,52,55] *Acanthamoeba* spp. *(A. castellanii* and *A. polyphaga)* have been isolated from some eye lesions[58-60] In some of the reported cases, steroids had been used in some form, and in other cases the cornea had been previously injured or had a long-standing history of *Herpes simplex* such as stromal herpetic keratitis.[34,42]

The histopathological features of ocular infections due to *Acanthamoeba* spp. are characterized by an acute or subacute necrotizing keratitis with an associated granulomatous inflammation. Diffuse, scant lymphocytic infiltrate with few multinucleated giant cells may be present. The conjunctival membrane may show a very intense congestion and acute and subacute inflammation. Amebic trophozoites and cysts may be identified by corneal biopsy (Figure 28). They are usually located in the corneal stroma and in front of the Descemet's membrane or deeper within the ciliary body and the uveal tissue. Iritis and uveitis may result by hematogenous dissemination rather than by direct invasion. The optic nerve has been reported as having an acute or subacute inflammatory reaction in experimental animals.[56]

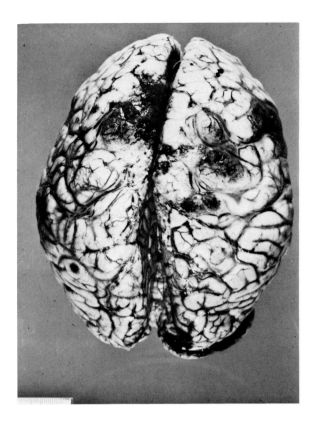

FIGURE 15.   GAE. Dorsal view. Scant purulent exudate is noted around cortical blood vessels, overlying the foci of hemorrhagic encephalomalacia. (From Presbyterian-University Hospital of Pittsburgh; PA-80-28. With permission.)

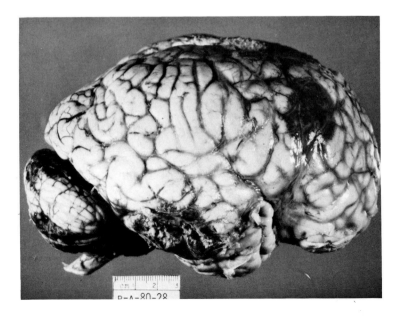

FIGURE 16.   GAE. Lateral view of cerebral hemispheres and cerebellum. Isolated areas of congestion are noted. (From Presbyterian-University Hospital of Pittsburgh; PA-80-28. With permission.)

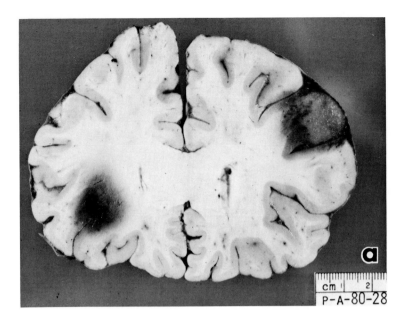

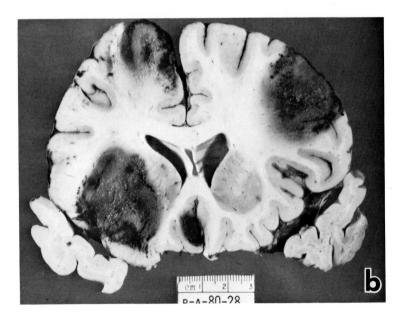

FIGURE 17.   (A, B, C, and D) GAE. Coronal sections of cerebral hemispheres showing multi-focal areas of hemorrhagic encephalomalacia involving cerebral crotex, subcortical white matter, and basal ganglia. (a) Level of frontal lobes; (b) level of olfactory nerves (arrows); (c) Level of mammillary bodies; (d) level of occipital lobes. (From Presbyterian-University Hospital of Pittsburgh; PA-80-28. With permission.)

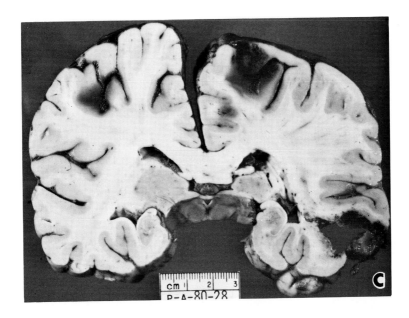

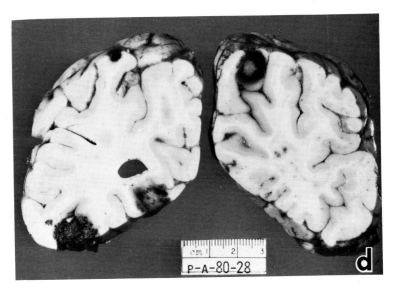

FIGURE 17 (continued)

The amebic trophozoites may be identified, also, in direct smears from the surface of the corneal ulcerations and may be cultured from this material. Ultrastructural studies of the corneal tissue may be done to demonstrate the fine structural details of amebic cysts and trophozoites (Figures 29 and 30). Ocular infections due to free-living amebas are not frequently found, probably because the normal immunological mechanisms are effective and because corneal damage with infection may be required for implantation of the amebic infection.

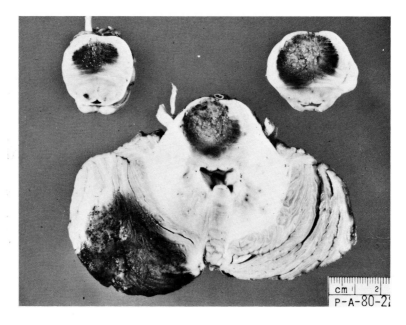

FIGURE 18.   GAE. Horizontal sections through the upper pons, midpons, and cerebellum. Large areas of necrosis and hemorrhages are seen. (From Presbyterian-University Hospital of Pittsburgh; PA-80-28. With permission.)

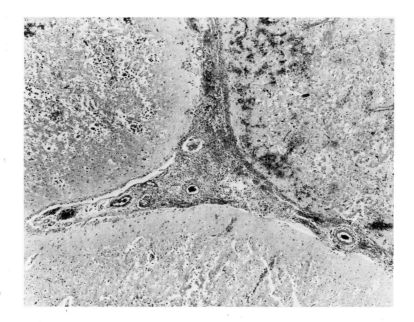

FIGURE 19.   GAE. Histologic section from one of the lesions shown in Figure 17. There are few multinucleated giant cells and scant lymphocytic infiltrate. Edema and gliosis are also present. (H & E; magnification × 250.) (From Presbyterian-University Hospital of Pittsburgh; PA-80-20. With permission.)

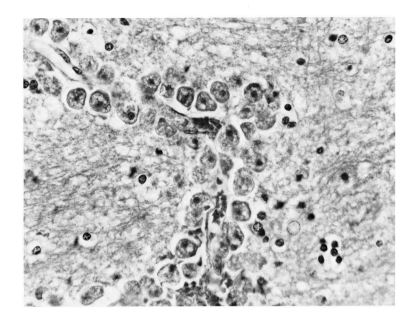

FIGURE 20.. GAE. Section of the cerebral cortex and subcortical white matter showing numerous amebic trophozoites around a blood vessel. (H & E; magnification × 250.) (From Presbyterian-University Hospital of Pittsburgh; PA-80-28. With permission.)

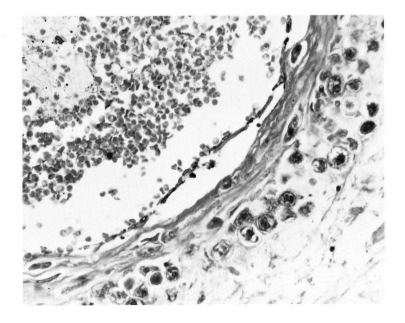

FIGURE 21. GAE. Cerebral cortex showing arteries containing numerous amebic trophozoites and cysts within their walls. Dr. G. Visvesvara was able to isolate and culture *A. castellanii* from autopsy tissue. (H & E; magnification × 200.) (From Presbyterian-University Hospital of Pittsburgh; PA-80-20. With permission.)

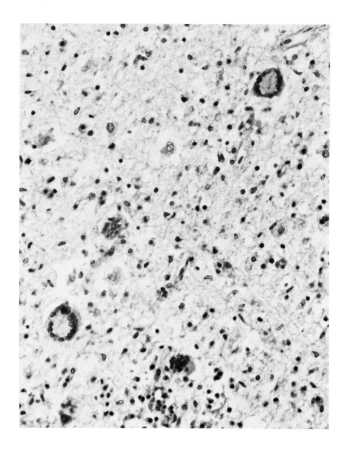

FIGURE 22.   GAE. Chronic granulomatous inflammatory reaction with multinucleated giant cells and reactive astrocytosis, subcortical white matter. (H & E; magnification × 800.) (From Presbyterian-University Hospital of Pittsburgh; PA-80-20. With permission.)

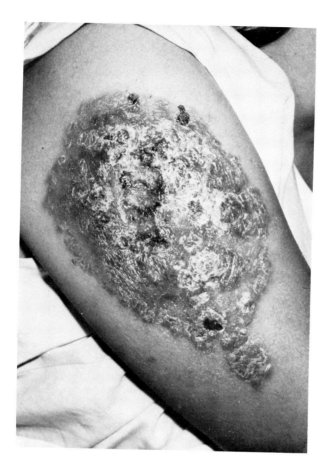

FIGURE 23. Chronic ulceration of skin on deltoid area. (Photo courtesy of Dr. John Gullet. (From Gullet, J. Mills, J., Hadley, K., et al., *Am. J. Med.*, 67, 891, 1979. With permission.)

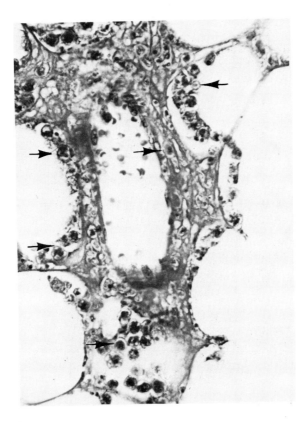

FIGURE 24.   Skin biopsy from one of the skin nodules showing a blood vessel within adipose tissue. Same case as Figure 17. Numerous amebic trophozoites can be seen around and within the blood vessel wall (arrows). (H & E; magnification × 250.) (From Presbyterian-University Hospital of Pittsburgh, PS-80-1350. With permission.)

FIGURE 25. Cut surface of lung. Large areas of consolidation involving upper and middle lobes of the lung may be found. (From Presbyterian-University Hospital of Pittsburgh; PA-80-20. With permission.)

FIGURE 26. Clusters of pseudohyphae of *Candida albicans,* within lung parenchyma. Amebic trophozoites and cysts are noted between numerous pseudohyphae, extending from the lumen of bronchi and alveolar walls. Cytomegalo virus was also found in the lungs. (Gomori-methenamine silver; magnification × 500.) (From Presbyterian-University Hospital of Pittsburgh; PA-80-20. With permission.)

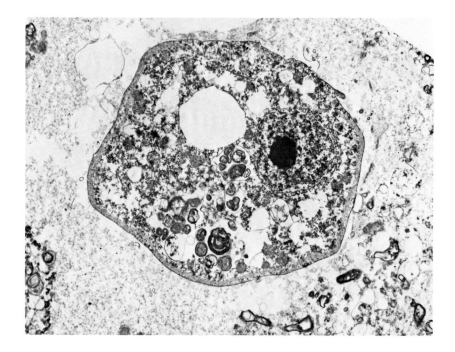

FIGURE 27.   Electron micrograph of one trophozoite of *Acanthamoeba castellanii* amidst necrotic CNS tissue. (Magnification × 7500.) (From Presbyterian-University Hospital of Pittsburgh; PA-80-28. With permission.)

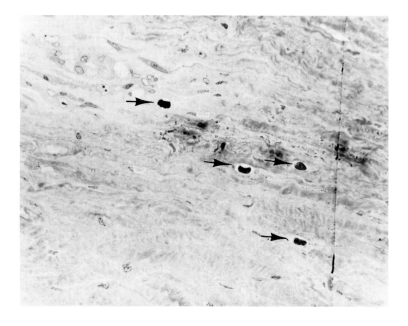

FIGURE 28.   Four acanthamoebic cysts within the corneal stroma (arrows). One-micron-thick, plastic embedded tissue. (Toluidine blue; magnification × 800.) (Biopsy of the cornea, courtesy of A. W. Dudley, Jr., M.D., S. Levin, J. T. McMalron, and D. Meisler, Cleveland Clinic.)

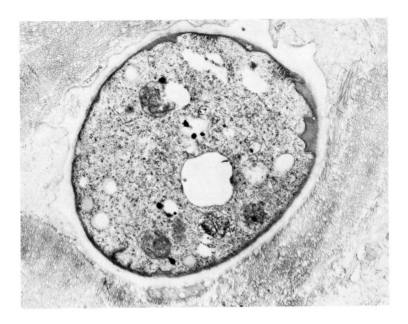

FIGURE 29. Electron micrograph of a double-layered acanthamoebic cyst surrounded by collagen fibrils from the cornea — case of a 31-year-old male with a right corneal ulceration treated initially with Ketaconazole and corneal transplant resulting in restored vision and no recurrence of infection. (Magnification × 10,000.) (Courtesy of Dr. A. W. Dudley, Jr., et al., Cleveland Clinic, 83-2985.)

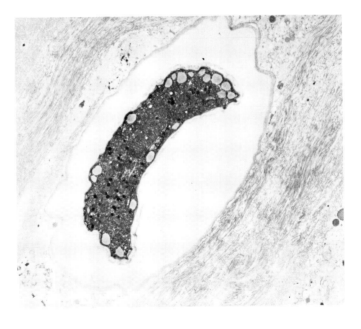

FIGURE 30. Electron micrograph from the same case as Figure 28 showing an elongated and a degenerated acanthamoebic cystic form, encompassed by a thin capsule and a clear space. (Magnification × 10,000.) (Courtesy of A. W. Dudley, Jr., M.D., et al. Cleveland Clinic.)

REFERENCES

1. Ashton, N. and Stamm, W., Amoebic infection of the eye. A pathological report, *Trans. Ophthalmol. Soc. U.K. (London)*, 95, 214, 1975.
2. Becker, G. L., Knep, S., Lance, K. P., and Kaufman, L., Amebic abscess of the brain, *Neurosurgery*, 6, 192, 1980.
3. Bhagwandeen, S. B., Carter, R. F., Naik, K. G., and Levitt, D., A case of *hartmannellid* amebic meningoencephalitis in Zambia, *Am. J. Clin. Pathol.*, 63, 483, 1975.
4. Borokovitz, D., Martinez, A. J., and Patterson, G. T., Osteomyelitis of bone graft of mandible with *Acanthamoeba castellanii* infection, *Hum. Pathol.*, 12, 573, 1981.
5. Bos, H. J., Volker-Dieben, H. J. M., and Kok-van Alphen, C. C., A case of *Acanthamoeba* keratitis in the Netherlands, *Trans. R. Soc. Trop. Med. Hyg.*, 75, 86, 1981.
6. Brant, H. and Peréz-Tamayo, R., Pathology of human amebiasis, *Hum. Pathol.*, 1, 351, 1970.
7. Butt, C. G., Primary amebic meningoencephalitis, *N. Engl. J. Med.*, 274, 1473, 1966.
8. Butt, C. G., Baro, C., and Knorr, R. W., *Naegleria* (sp.) identified in amebic encephalitis, *Am. J. Clin. Pathol.*, 50, 568, 1968.
9. Cain, A. R., Wiley, P. F., Brownell, D. B., and Warhurst, D. C., A fatal case of primary amebic meningoencephalitis, *Arch. Dis. Childhood*, 56, 140, 1981.
10. Callicott, J. F., Jr., Amebic meningoencephalitis due to free-living amebas of the *Hartmannella (Acanthamoeba)-Naegleria* group, *Am. J. Clin. Pathol.*, 49, 84, 1968.
11. Callicott, J. H., Nelson, E. C., Jones, M. M., dos Santos, J. G., Utz, J. P., Duma, R. J., and Morrison, J. V., Meningoencephalitis due to pathogenic free-living amebas, *J. A. M. A.*, 206, 579, 1968.
12. Carter, R. F., Primary amebic meningoencephalitis: clinical, pathological and epidemiological feature of six fatal cases, *J. Pathol. Bacteriol.*, 96, 1, 1968.
13. Carter, R. F., Description of a *Naegleria* sp. isolated from two cases of primary amebic meningoencephalitis, and of the experimental pathological change induced by it, *J. Pathol.*, 100, 217, 1970.
14. Carter, R.F., Primary amebic meningoencephalitis: an appraisal of present knowledge, *Trans. R. Soc. Trop. Med. Hyg.*, 66, 193, 1972.
15. Carter, R. F., Cullity, G. J., Ojeda, V. J., Silberstein, P., and Willaert, E., A fatal case of meningoencephalitis due to a free-living ameoba of uncertain identity-probably *Acanthamoeba* sp., *Pathology*, 13, 51, 1981.
16. Cleland, P. G., Lawande, R. V., Onyemelukwe, G., and Whittle, H. C., Chronic amebic meningoencephalitis, *Arch. Neurol.*, 39, 56, 1982.
17. Cox, E. C. Amebic meningoencephalitis caused by *Acanthamoeba* species in a four month old child, *J. S. C. Med. Assoc.*, 76, (10), 459, 1980.
18. Culbertson, C. G., Pathogenic *Acanthamoeba (Hartmannella)*, *Am. J. Clin. Pathol.*, 35, 195, 1961.
19. Culbertson, C. G., Ensminger, P. W., and Overton, W. M., *Hartmannella (Acanthamoeba):* experimental chronic granulomatous brain infections produced by new isolates of low virulence, *Am. J. Clin. Pathol.*, 46, 305, 1966.
20. Darby, C. P., Conradi, S. E., Holbrook, T. W., and Chatellier, C., Primary amebic meningoencephalitis, *Am. J. Childhood Dis.*, 133, 1025, 1979.
21. Duma, R. J., Ferrell, H .W., Nelson, E. C., and Jones, M. M., Primary amebic meningoencephalitis, *N. Engl. J. Med.*, 281, 1315, 1969.
22. Duma, R. J., Rosenblum, W. I., McGehee, R. F., Jones, M. M, and Nelson, E. C., Primary amoebic meningoencephalitis caused by *Naegleria*. Two new cases. Reponse to amphotericin B and a review, *Ann. Intern. Med.*, 74, 923, 1971.
23. Duma, R. J., Helwig, W. B., and Martinez, A. J., Meningoencephalitis and brain abscess due to a free-living amoeba, *Ann. Intern. Med.*, 88, 468, 1978.
24. Flexner, S., Amoebae in an abscess of the jaw, *John Hopkins Hosp. Bull.*, 3, 104, 1892.
25. Fowler, M. and Carter, R. F., Acute pyogenic meningitis probably due to *Acanthamoeba* spp.: a preliminary report, *Br. Med. J.*, 2, 740, 1965.
26. Grunnet, M. L., Cannon, G. H., and Kushner, J. P., Fulminant amebic meningoencephalitis due to *Acanthamoeba*, *Neurology*, 31, 174, 1981.
27. Gullett, J., Mills, J., Hadley, K., Podemski, B., Pitts, L., and Gelber, R., Disseminated granulomatous *Acanthamoeba* infection presenting as an unusual skin lesion, *Am. J. Med.*, 67, 891, 1979.
28. Hamburg, A. and De Jonckheere, J. F., Amoebic keratitis, *Ophthalmologica (Basel)*, 181, 74, 1980.
29. Hecht, R. H., Cohen, A., Stoner, J., and Irwin, C., Primary amebic meningoencephalitis in California, *Calif. Med.*, 117, 69, 1972.
30. Hermanne, J., Jadin, J. B., and Martin, J. J., Meningoencephalite amibienne primitive en Belgique. A propos de deux cas, *Ann. Pediatr.*, 19, 425, 1972.

31. Hoffmann, E. O., Garcia, C., Lunseth, J., McGarry, P., and Coover, J., A case of primary amebic meningoencephalitis. Light and electron microscope and immunohistologic studies, *Am. J. Trop. Med. Hyg.*, 27, 29, 1978.

32. Jager, B. V. and Stamm, W. P., Brain abscesses caused by free-living amoeba probably of the genus *Hartmannella* in a patient with Hodgkin's disease, *Lancet*, 2, 1343, 1972.

33. Jones, B. R., McGill, J. I., and Steele, A. D. McG., Recurrent suppurative kerato-uveitis with loss of eye due to infection by *Acanthamoeba castellani*, *Trans. Ophthalmol. Soc. U.K. (London)*, 95, 210, 1975.

34. Jones, D. B., Visvesvara, G. S., and Robinson, N. M., *Acanthamoeba polyphaga* keratitis and *Acanthamoeba* uveitis associated with fatal meningoencephalitis, *Trans. Ophthalmol. Soc. U.K. (London)*, 95, 221, 1975.

35. Kenney, M., The micro-kolmer complement fixation test in routine screening for soil ameba infection, *Health Lab. Sci.*, 8, 5, 1971.

36. Kernohan, J., Magath, T.B., and Schloss, G. T., Granuloma of brain probably due to *Endolimax williamsi (Iödamoeba bütschlii)*, *Arch. Pathol.*, 70, 576, 1960.

37. Key, S. N., Green, W. R., Willaert, E., and Stevens, A., Keratitis due to *Acanthamoeba castellanii*. A clinicopathological case report, *Arch. Ophthalmol.*, 98, 475, 1980.

38. Lawande, R. V., MacFarlane, J. T., Weir, W. R. C., and Awunor-Renner, C., A case of primary amebic meningoencephalitis in a Nigerian farmer, *Am. J. Trop. Med. Hyg.*, 29(1), 21, 1980.

39. Lengy, J., Jakovlzevich, R., and Tolis, B., Recovery of a hartmanelloid amoeba from a purulent ear discharge, *Trop. Dis. Bull.*, 68, 818, 1971.

40. Lombardo, L., Alonso, P., Arroyo, L. S., Brandt, H., and Mateos, J. H., Cerebral amebiasis: report of 17 cases, *J. Neurosurg.*, 21, 704, 1964.

41. Lund, O. E., Stefani, F. H., and Dechant, W., Amoebic keratitis: a clinicopathological case report, *Br. J. Ophthalmol.*, 62, 373, 1978.

42. Ma, P., Willaert, E., Ujuechter, K. B., and Stevens, A. R., A case of keratitis due to *Acanthamoeba* in New York, New York, and features of 10 cases, *J. Infect. Dis.*, 143, 662, 1981.

43. Markowitz, S. M., Martinez, A.J., Duma, R. J., and Shiel, F. O. M., Myocarditis associated with primary amebic *(Naegleria)* meningoencephalitis, *Am. J. Clin. Pathol.*, 62, 619, 1974.

44. Martinez, A. J., dos Santos, J. G., Nelson, E. C., Stamm, W. P., and Willaert, E., Primary amebic meningoencephalitis, *Pathology Annual*, Part 2, Sommers, S. C. and Rosen, P. P., Eds., 12, 225, 1977.

45. Martinez, A. J., Sotelo-Avila, C., Garcia-Tamayo, J., Takano Morón, J., Willaert, E., and Stamm, W. P., Meningoencephalitis due to *Acanthamoeba* sp. Pathogenesis and clinicopathological study, *Acta Neuropathol. (Berl.)*, 37, 183, 1977.

46. Martinez, A. J., Garcia, C. A., Halks-Miller, M., and Arce-Vela, R., Granulomatous amebic encepahlitis presenting as a cerebral mass lesion, *Acta Neuropathol. (Berl.)*, 51, 85, 1980.

47. Martinez, A. J., Is acanthamoebic encephalitis an opportunistic infection?, *Neurology (New York)*, 30, 567, 1980.

48. Martinez, A. J., Sotelo-Avila, C., Alcalá, H., and Willaert, E., Granulomatous encephalitis, intracranial arteritis and mycotic aneurysm due to a free-living ameba, *Acta Neuropathol. (Berl.)*, 49, 7, 1980.

49. Martinez, A. J., The spectrum of free-living amebic infections of the brain (abstr.), 9th Int. Congr. Neuropathology, Vienna, Austria, September 1982.

50. Martinez, A. J., Acanthamoebiasis and immunosuppression. Case report, *J. Neuropathol. Exp. Neurol.*, 41, 548, 1982.

51. Nagington, J., Watson, P. G., Playfair, T. J., McGill, J., Jones, B. R., and Steele, A. D. McG., Amoebic infection of the eye, *Lancet*, 2, 1537, 1974.

52. Nagington, J., Isolation of amoebae from eye infection in England, *Trans. Ophthalmol. Soc. U.K. (London)*, 95, 207, 1975.

53. Patras, D. and Andujar, J. J., Meningoencephalitis due to *Hartmannella (Acanthamoeba)*, *Am. J. Clin. Pathol.*, 46, 226, 1966.

54. Robert, V. B. and Rorke, L. B., Primary amebic encephalitis probably from *Acanthamoeba*, *Ann. Intern. Med.*, 79, 174, 1973.

55. Samples, J. R., Binder, P. S., Luibal, F. L., Font, R. L., Visvesvara, G. S., and Peter, C. R., *Acanthamoeba* keratitis possibly acquired from a hot tub, *Arch. Ophthalmol.*, 102, 707, 1984.

56. Schlaegel, T. F. and Culbertson, C. G., Experimental *Hartmannella* optic neuritis and uveitis, *Ann. Ophthalmol.*, p. 103, 1972.

57. Stein, A. and Kazan, A., Brain abscess due to *Entameba histolytica*, *J. Neuropathol. Exp. Neurol.*, 1, 34, 1942.

58. Visvesvara, G. S., Jones, D. B., and Robinson, N. M., Isolation, identification, and biological characterization of *Acanthamoeba polyphaga* from a human eye, *Am. J. Trop. Med. Hyg.*, 24, 784, 1975.

59. Warhurst, D. C., Stamm, W.P., and Phillips, E. A., *Acanthamoeba* from a new case of corneal ulcer, *Trans. R. Soc. Trop. Med. Hyg.*, 70, 279, 1976.

60. Watson, P. G., Amoebic infection of the eye, *Trans. Ophthalmol. Soc. U.K. (London)*, 95, 204, 1975.
61. Wessel, H. B., Hubbard, J., Martinez, A. J., and Willaert, E., Granulomatous amebic encephalitis (GAE) with prolonged clinical course: CT scan findings, diagnosis by brain biopsy, and effect of treatment, *Neurology (Minneap.)*, 30, 442, 1980.

Chapter 7

LABORATORY DIAGNOSIS: PAM AND GAE — CEREBROSPINAL FLUID EXAMINATION

I. PRIMARY AMEBIC MENINGOENCEPHALITIS (PAM)

The most important laboratory procedure for the diagnosis of primary amebic meningoencephalitis (PAM) due to *Naegleria fowleri* is the direct microscopic examination of CSF as originally used and recommended by Dr. J. G. dos Santos.[42] A hanging drop or a simple wet-mount examination using CSF on a glass slide may demonstrate the limax motile amebas. *N. fowleri* are usually 8 to 15 μm in largest dimensions.[8] The diagnosis can be made in the emergency room with CSF obtained by lumbar or cisternal puncture.[43] This is done by taking one or two drops of the sediment from the CSF and placing them onto a glass slide at room temperature (about 25°C). It is preferable to centrifuge at low speeds to concentrate the trophozoites (about 150 × g for 15 min). After centrifuging the supernatant may be used for chemical analysis and the sediment may be used for microscopic examination by simply lowering the condenser of the microscope or constricting the diaphragm in order to enhance the contrast and reduce the illumination. Phase contrast illumination may be used. Nomarski optics as well as vital stains using Congo red, Janus Green B, methylene blue, or Lugol solution may be used or the indirect immunofluorescence technique.[58,68] Amebas can be seen moving in the CSF under the microscope.[12,20,64]

Culture of the CSF is necessary for complete diagnosis.[15,22,26] In general, the CSF, except for the hemorrhagic component, is indistinguishable from that obtained from patients having bacterial or pyogenic meningitis or meningoencephalitis. The color of the CSF varies from grayish to yellowish-white tinged with red and with a hazy to opaque appearance. Table 1 compiled by Dr. J. G. dos Santos summarized the findings of the 16 cases from Virginia. The red blood cell count varies with the stage of the disease, being low initially and increasing as the disease progresses. The red blood cell count has been as low as 250/mm³ and as high as 24,600/mm. Because PAM due to *N. fowleri* is a hemorrhagic and necrotizing meningoencephalitis, the hemorrhagic and proteinaceous components are very conspicuous. The white blood cell count in the CSF in PAM due to *Naegleria* also varied from as low as 300 mm³ to as high as 26,000 mm³ with a differential similar to bacterial leptomeningitis or pyogenic or suppurative meningoencephalitis, such as predominance of polymorphonuclear leukocytes, few lymphocytes and perhaps some monocytes, and, of course, no bacteria present in the CSF. Antibiotics, of course, may produce a sterile CSF even in cases of bacterial meningoencephalitis.

The differential white blood cell count, in percent, may range as follows:

Polymorphonuclear leukocytes	52— 99%
Lymphocytes	1—48%
Monocytes	1—2%

A distinctive feature of *N. fowleri* is its ability to convert into a flagellate form which can be induced by dilution of the culture with water. In CSF, the amebas are easily confused with leukocytes but they can be identified by the typical central, spherical nucleolus characteristics of the limax amebas.[50,57]

By phase contrast illumination or simply by microscopic examination of the CSF,

Table 1

FATAL CASES OF PAM DIAGNOSED AT THE MEDICAL COLLEGE OF VIRGINIA FROM 1937 THROUGH 1969

FATAL CASES OF PRIMARY AMEBIC MENINGOENCEPHALITIS DIAGNOSED AT THE MEDICAL COLLEGE OF VIRGINIA from 1937 through 1969 (compiled by J.G. dosSantos, Neto, M.D.)

Case No.	Hospital	Name	Chart No.	Age	Sex	Onset	Adm.	Death	Swimming in fresh water	Cells (cu. mm)	CEREBROSPINAL FLUID						Autopsy
											Diff.	Protein (mg%)	Sugar (mg%)	Bacteria Smear-Cult.	Amoeba in Fluid	Amoeba Culture	
O 1	St. Elizabeth	EVF	B-4564 inactive	13	F	7-15-37	7-19-37	7-19-37	Yes Lake A	RBC:2; WBC:26000	mostly polys			cult. neg. no smear			A2155-done at MCV Positive for amoeba
O 2	MCV	MLR	B-09-43-92	12	F	9-10-50	9-12-50	9-13-50	Yes Lake A	WBC: 1650	P-69% L-32%	450	44 (BS 190)	negative			A6467 Positive for amoeba
△ 3	MCV	JOM	B-11-77-61	23	M	7-16-51	7-18-51	7-20-51	Yes Lake A	1452	P-78% L-22%	600	90 / 13	negative			A6820 Positive for amoeba
△ 4	MCV	ERM	B-11-79-98	14	F	7-24-51	7-25-51	7-25-51	Yes Lake B	2970	P-98% L-2%	970	20	negative			A6829 Positive for amoeba
● 5	MCV	EWM	B-11-82-14	14	M	7-24-51	7-25-51	7-26-51	Yes Lake A	391 / 1768	P-79% L-21% / P-72% L-28%	150 / 550	90 / 66	negative			A6832 Positive for amoeba
△ 6	MCV	JLB	B-02-36-56	8	M	6-29-52	7-2-52	7-3-52	Yes Lake B	1490	P-95% L-5%	280		negative			A7235 Positive for amoeba
△ 7	MCV	DLT	B-14-65-95	21	F	7-3-52	7-5-52	7-7-52	Yes Lake B	974	P-521 L-453	75 / 600	18 at death	negative			A7238 Positive for amoeba
△ 8	MCV	PTK	B-14-67-74	17	F	7-6-52	7-8-52	7-8-52	Yes Lake B	21400	P-94%	350	15	negative			A7240 Positive for amoeba
△ 9	St. Luke's	NB	97-399	25	F	7-3-52	7-8-52	7-9-52	Yes Lake B	3300	P-97% L-3%	500	<10	negative			A7242-done at MCV Positive for amoeba
O10	MCV	HTS, JI	B-26-68-78	13	M	7-23-57	7-28-57	7-29-57	Yes Lake B	5390	P-95% L-5%	400	12	negative			A9802 Positive for amoeba
O11	MCV	MLB	B-26-83-76	11	M	8-15-57	8-17-57	8-18-57	Yes	12750	P-12190 L-560	4+	<7	negative			A9828 Positive for amoeba

▲12	MCV	WHK	5-17-83-70	14	M	7-28-66	7-29-66	7-30-66	Yes Lake B	RBC: 250 WBC: 4900	P-96 L-2 M-2	90	20 (BS 139)	negative	yes	yes*	A349-66 Positive for amoeba
□13	MCV	CLJ	9-10-49-50	27	M	7-28-67	7-31-67	8-1-67	Yes Lake C	RBC: 24600 WBC: 9900	P-94 L-3 M-2	QNS		negative	yes	yes*	A338-67 Positive for amoeba by smear and culture*
■14	MCV	LJL	5-25-19-16	15	F	8-11-68	8-13-68	8-15-68	Yes Lake A	RBC: 310 WBC: 300	P-83 L-10 M-1 10 amoe. /100WBC	580	12 (BS 235)	negative	yes 118/cu. mm.	yes*	A339-68 Positive for amoeba by smear and culture*
◇15	MCV	TRY	5-28-69-77	14	M	7-9-69	7-10-69	7-14-69	Yes Lake B	RBC: 530 WBC 1540	P-70 L-28 M-2	QNS	6 (BS 178)	negative	yes 4/100 RBC smear	yes*	A 250-69 Amoeba seen in slides and cultured.*
●16	MCV	WDM	5-14-70-43	24	M	7-10-69	7-12-69	7-15-69	Yes Lake B	RBC: 500 WBC 9075	P-99 L-1	400	70 (BS 208)	negative	yes 6/100 RBC smear	yes*	A 251-69 Amoeba seen in slides and cultured*

O Cases diagnosed in retrospect by review of 16,174 autopsies (1920-1968).

● Case witnessed but only diagnosed in retrospect by reviewing pathology slides.

△ Cases witnessed but only diagnosed in retrospect by reviewing pathology slides. (Reported by Dr. J. Callicott, Am. J. Clin. Path., 49:1, 84-91, 1968)

▲ First case, diagnosed postmorten by Dr. J. Callicott (Dept. of Path.) (Reported Am. J. Clin. Path., 49:1, 84-91, 1968)

□ Case diagnosed during life by the demonstration of amoeba when performing CSF red cell count reported in JAMA Oct. 14, 1968.

■ Case suspected clinically and diagnosed by Dr. Richard Duma (Dept. of Medicine, section of Infectious diseases) Reported in Mortality and Morbidity Report of CDC, 17:36, 330, Sept 7, 1968, and New Eng. J. Med., 281:24, 1315-1323, Dec. 11, 1969

* Organism cultured by Dr. E. C. Nelson, Professor of Parasitology, MCV.

◇ Case diagnosed by Dr. Cary Suter, Chairman, Division of Neurology, Department of Medicine, MCV.

◆ Case diagnosed by Dr. Maurice Nottingham, Richmond Memorial Hospital, and transferred to the Infectious Disease Section, Dept. of Medicine, MCV.

ᵃ Lake A is Lake Moore; Lake B is Lake Chester; Lake C is Lake Manchester.

Compiled by J. G. dos Santos, Neto, M.D. With permission.

the trophozoites can be seen to contain a single central or eccentric nucleus with a conspicuous karyosome and characteristic blunt pseudopodia (lobopodia). Smears of CSF (centrifuged if the white blood cell count is low) offer the advantage of making a permanent preparation which can be kept indefinitely in the files. In addition to showing the different white blood cells, the presence of free-living amebas may be easily demonstrated. Wright or Giemsa stain are recommended by dos Santos.[42] Visvesvara recommended Wheatley's trichrome or Heindenhain's iron hematoxylin.[43] With Giemsa or Wright stains the amebas have a delicately refractile sky-blue cytoplasm with slightly pink nuclei, while lymphocytes and monocytes have large nuclei and scant cytoplasm. The Wright or Giemsa-stained smears play an important role in the diagnosis because cases of PAM, when seen for the first time, are often mistaken for bacterial leptomeningitis. As a result, amebas are neither searched for nor cultured. Yet, Gram's stained and Wright or Giemsa-stained smears are usually made and available and, while the amebas are poorly defined and passed off as degenerate cells on the Gram's stained smears, they stand out in the smears when the white blood cell differential count is performed (Figures 1 and 2).

Acridine orange stain using an ultraviolet light may be useful in the diagnosis because the amebic trophozoites stain a brick-red color with a pale green nucleus in contrast to the leukocytes which stain bright green.[44] The cysts stain with a uniform bright brick-red color.

Greenstein's five-dye stain has also been used for axenic cultures of *Acanthamoeba* spp. and *N. fowleri*. The karyosome and the chromatin of the nucleus are stained bright red or brick red.[54]

The protein concentration in the CSF is elevated; values may range from a low of 75 mg/ℓ to a high value of 970 mg/dℓ. The protein concentration in the CSF is elevated with high values of gammaglobulin. The glucose value is very low as in bacterial or purulent meningitis. The CSF may be under pressure, therefore, the physician performing the lumbar or cisternal puncture should be aware of this possibility to avoid cerebellar tonsillar herniation during the withdrawal of the CSF. Measurement of CSF pressure should be recorded during the procedure. This examination is important because cases of PAM, when seen for the first time, are often mistaken for pyogenic or purulent meningitis.[25]

II. GRANULOMATOUS AMEBIC ENCEPHALITIS (GAE)

The CSF in cases of GAE due to *Acanthamoeba* sp. may have a predominance of lymphocytes, but until now no diagnosis of GAE has been reported using the CSF while the patient was still alive. The glucose concentrations are as low as in bacterial (purulent) leptomeningitis. The CSF should be transparent and the protein may be slightly elevated. It is imperative that serum blood sugar levels be determined simultaneously for diagnostic comparison with those of CSF. Measurements of CSF pressure should also be recorded. Elevations of more than 500 mm of H_2O have been reported[20] (Table 2).

Gram-stained smears and bacterial cultures should be negative in all cases. The amebic trophozoites should be demonstrated well with Giemsa or Wright stains similar to the cases of PAM. The search for motile amebas with their typical movements is of paramount importance and should be achieved in the same way as already explained by PAM. Trophozoites of *Acanthamoeba* spp. are usually between 15 to 25 μm.[50] Phase or darkfield microscopy may be helpful for visualization of the trophozoites as with PAM.

Brain biopsy and frozen sections may demonstrate the necrosis of cerebral tissue, associated with inflammatory reaction composed mainly of lymphocytes, multinu-

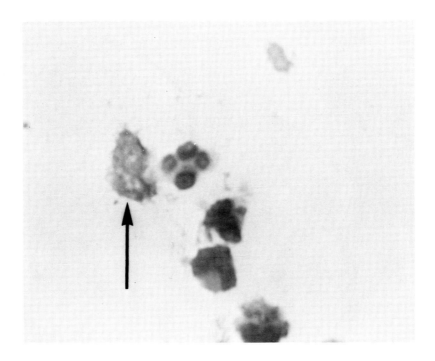

FIGURE 1. Trophozoites (arrow) of *Naegleria fowleri* from the CSF demonstrating the characteristic centrally placed nuclei and the abundant cytoplasm. (Giemsa; magnification × 600.)

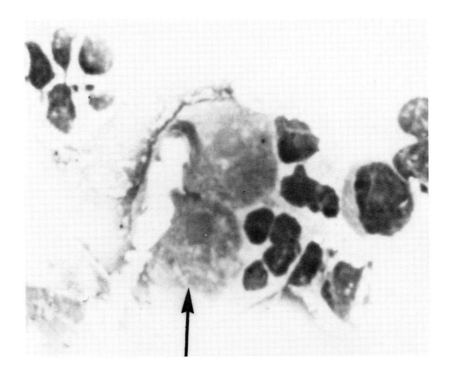

FIGURE 2. Trophozoites of *Naegleria fowleri* from the CSF (arrow). The characteristic cytological features of the trophozoites are not well demonstrated. (Gram; magnification × 600.)

Table 2
CSF FINDINGS IN GRANULOMATOUS AMEBIC ENCEPHALITIS

Case #	Place	Age	Sex	Race	WBC/CUMM (cells/μl)	Polys (%)	Lymphs (%)	Glucose (mg/dl)	Protein (mg/dl)
1	Arizona	6	F	W	204	13	87	35	214
2	Australia	2½	F	W	—	?	63	60	33
3	California	24	F	W	66	2	98	52	70
4	Honduras	56	M	W	65	—	—	160	96
5	Idaho	57	F	W	7	75	—	84	112
6	India	30	M		—	—	250	74	300
7	New Orleans	33	F	W	67	—	—	48	111
8	New Orleans	56	M	B	?	0	30	50	60
9	New York	27	M	W	51	32	19	51	92
10	Peru	20	M	W	750	0	94	52,95	54,97
11	Peru	22	M	W	14	0	100	67	?
12	Peru	42	M	W	70	30	?	?	?
13	Peru	9	F	W	68	26	72	46	54
14	Philadelphia	58	M	W	520	9	91	56	500
15	Pittsburgh	2½	M	W	—	—	22	38	62
16	Pittsburgh	38	M	W	1	—	1	78	110
17	Texas	7	M	B	106	20	80	66	31
18	Texas	2½	F		110	—	100	17	High[a]
19	Virginia	47	F	B	110	5	95	240	84
20	Zambia	Adult		B	290,220	62,45	38,55	16,30	85,146

[a] Too high to read.

cleated giant cells, plasma cells, and few polymorphonuclear leukocytes. The characteristic amebic trophozoites and cysts should also be found perivascularly or within the necrotic tissue.

III. CULTURE, ISOLATION, AND IDENTIFICATION OF FREE-LIVING AMEBAS

The reproduction and growth of free-living amebas in culture media has provided a means to study these protozoa in a controlled environment subjected to experimental manipulation with desirable conditions and with specific nutritional requirements. Culture and isolation of free-living amebas is important for definitive diagnosis and taxonomical classification of cases of PAM and GAE.[3,5,15-18,66]

The observations made under experimental conditions may be similar to those in vivo, but valuable information may be obtained with axenic or nonaxenic cultures. Cultivation normally permits the production of larger amounts of trophozoites and cysts that can be used for morphological, biochemical, immunological, and serological studies. Some culture media are particularly useful for physiological, biochemical, and, especially, chemotherapeutic studies.[26,39-41,55,56]

Suitable materials for culture and isolation are

1. Cerebrospinal fluid (CSF)
2. CNS tissues
3. Nasal or oropharyngeal secretions
4. Skin lesions, pulmonary or ocular tissues
5. Water from: cooling towers, condensers, dehumidifiers, etc.
6. Mud, sand, dust, soil

Some body secretions (gastrointestinal fluid) may be contaminated with free-living amebas giving the impression of amebic infection. Thus far, no other cytologic preparation (pleural fluids, ascitic fluid, etc.) has been reported containing free-living amebas.[37]

The material received by the pathologist, cell biologist, or laboratory technician should be placed immediately for cultivation in the appropriate medium.[65] Another portion may be used for inoculation in tissue cultures or in the experimental animal. The specimens should be kept at room temperature and should not be frozen or refrigerated.[64] Some of the free-living amebas are virulent and potentially pathogenic for humans. Therefore, the technical personnel working with free-living amebas should take appropriate precautions as in any other infectious and contagious disease. Surgical masks, gloves, and operating gowns are mandatory and the specimen should be handled in a biological safety cabinet. Flagellation experiments may be done placing a drop of CSF or culture medium in about 1 mℓ of distilled water and observing this preparation under the microscope. Usually *Naegleria fowleri* may develop between two to nine flagella when placed in distilled water. This test will classify the free-living ameba as an amebo-flagellate.

IV. MEDIA AND PROCEDURES FOR CULTURE AND ISOLATION

The isolation of free-living amebas and their preservation in the culture media is of paramount importance for the taxonomical classification and for final identification of any strain of pathogenic and nonpathogenic free-living amebas.[21,25,48,49]

There are several types of culture media: liquid, solid, or semisolid. The choice of medium depends on the purpose, for example, isolation only, preservation, and continuation of a strain or for the characterization of the biochemical and antigenic properties of a specific strain.

Free-living amebas in axenic culture may loose infectivity or pathogenicity for experimental animals. The virulence could be restored by passages into mice or adding bacteria supplement to the medium. Different concentrations and types of sera from vertebrates may be used to enhance the growth of the amebas.[36] Antibiotics may be added (single or in combinations) such as nystatin, streptomycin, penicillin, Terramycin®, Polymyxin B®, or 5-fluorocytosine at different concentrations. In this way an overgrowth of undesirable fungi and bacteria may be inhibited. In general, the solid media uses agar at 1.5 to 3% in addition to live or killed bacteria such as *Escherichia coli, Aerobacter aerogenes (Klebsiella pneumoniae)* or *Proteus mirabilis.*[11]

The temperature and pH of the culture media have a direct influence on the growth of free-living amebas. Virus-like and bacterium-like structures have been reported when *Naegleria gruberi* grows at 37°C or at 21°C.[55,56] The pathogenic strains of *N. flowerli* are thermophilic and grow better at higher temperatures (up to 45°C), while *Acanthamoeba* spp. and the nonpathogenic *N. gruberi* grow better at room temperature (between 25 to 30°C).[9,34,35,59,60] Some media contains alcohol-killed bacteria and others have been devised using alcohol-killed bacteria supplemented with folic acid.[45]

The following procedures can be used to isolate the free-living ameba from the biological specimen or from natural sources, such as water, mud, or even the air.

The procedure recommended by Drs. E. C. Nelson and Muriel M. Jones consists of using flat-sided vessels (Leighton, Fisher, or agar slant) with about 10 mℓ of medium (Figure 3). After inoculation the flasks are incubated at a slant (about 45°C angle) to induce crawling of the amebas up the wall. A celluloid strip may be introduced into the medium for amebic adhesion. This strip may be fixed and stained for demonstration of the trophozoites. Examination for growth can be made by direct viewing through the wall on the slanted stage of an inverted or conventional microscope (Figures 4 to 7).

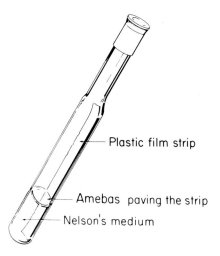

FIGURE 3.   Flask containing the Nelson's medium, a celluloid strip (showing where the amebas stick), and the culture.

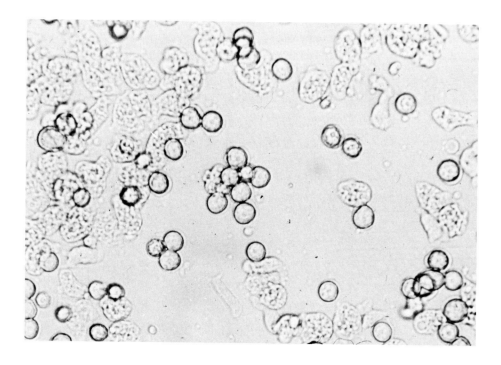

FIGURE 4.   Examination of the culture medium through the microscope. Abundant trophozoites and cysts of *Naegleria fowleri* are seen. Lee strain 12th day in axenic culture. (Magnification × 400.) (Courtesy of Dr. E. C. Nelson.)

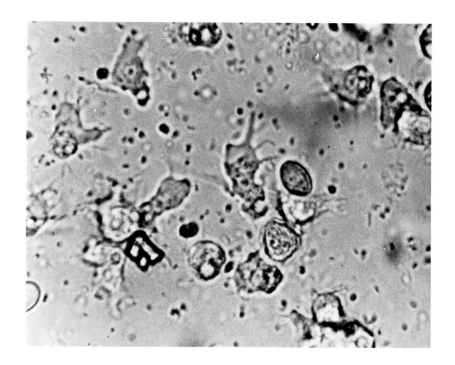

FIGURE 5.   *Acanthamoeba castellanii* trophozoites and cysts; 3 days in agar culture. (Magnification × 400.) (Courtesy of Dr. E. C. Nelson.)

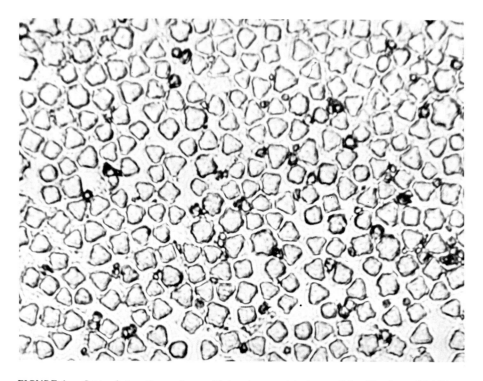

FIGURE 6.   Cysts of *Acanthamoeba* spp. 20 days in nonnutrient agar. (Magnification × 400.) (Courtesy of Dr. E. C. Nelson.)

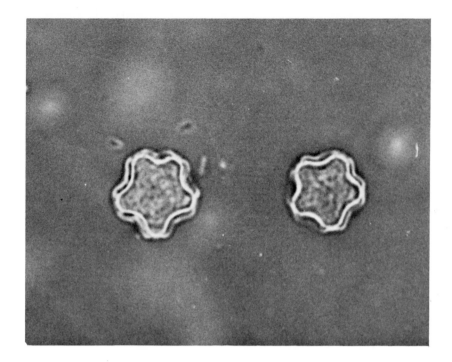

FIGURE 7.   Cysts of *Acanthamoeba* spp. on agar surface. Agar slant culture. (Magnification × 800.) (Courtesy of Dr. E. C. Nelson.)

The specimens which contain free-living amebas may be cultured in nonnutrient agar plates (solid medium) in the Nelson's liquid or axenic medium for *N. fowleri,* in the media of Červa, Page, Chang, Willaert, or modifications of these such as the one containing serum-casein-glucose-yeast extract (SCGYEM).[49]

A. Nonnutrient Agar

The nonnutrient agar are mainly used for the culture and isolation of pathogenic free-living amebas; their formula is as follows:

Agar	1.5 g
Page's ameba saline	100 m*l*

The agar is dissolved in the saline with heat and sterilized at 15 lb/in.[2] pressure for 15 min. It is cooled to 60°C and aseptically poured into plastic petri dishes, using 20m*l* for 100 × 15 mm dishes or 5 m*l* for 16 × 15 mm dishes. After the agar gels, store the plates in the refrigerator in canisters at 4°C. Plates may be kept in the refrigerator for about 3 months.

The formula of Page's ameba saline is

Sodium chloride (NaCl)	120 mg
Magnesium sulfate (MgSO$_4$.7H$_2$O)	4 mg
Calcium chloride (CaCl$_2$.2H$_2$O)	4 mg
Disodium hydrogene phosphate (Na$_2$HPO$_4$)	142 mg
Potassium dihydrogen phosphate (KH$_2$PO$_4$)	136 mg
Distilled water to	1 *l*

The chemicals are dissolved in the water and sterilized by autoclaving at 15 lb/in.[2] for 15 min. The solution may be stored in the refrigerator for up to 6 months at pH 6.8. The saline solution of Page[45] is a modification of those of Neff[46] and Band.[3]

B. Nelson's Axenic Medium for *N. fowleri*

Panmede (ox liver digest)	1.0 g
Glucose	1.0 g
Ameba Page's saline to	1000 mℓ

1. Ingredients dissolved in Page's ameba saline (see nonnutrient agar plates).
2. Dispense in 16 × 125 mm screw-cap tubes, 10 mℓ per tube
3. Autoclaved for 15 min
4. 0.2 mℓ of heat-inactivated fetal calf serum added to each tube before inoculating the medium with amebas

The shelf life at room temperature is several weeks. To transfer from one tube to another, gently swirl or stir the suspension. About four (4) drops of the medium may be obtained and subcultured. For nasal instillation in the mouse all fluid but about 1 mℓ can be removed, and wash the slope down with fluid. A Pasteur pipette or a dropper can be used.

C. Červa's Medium for *N. fowleri*

The first pathogenic strain of *N. fowleri* was isolated by Lubor Červa in June 1968. Červa used petri dishes with 1% Difco Bacto-Agar® which had been coated with a suspension of a thermally killed culture of *Aerobacter aerogenes*.[11] The petri dishes were placed upside down in plastic bags hermetically closed and incubated at 37°C. In addition, Červa developed a liquid axenic medium composed of 2% Bacto-Casitone® (Difco) in distilled water with 10% of fresh horse serum and 0.5% of sodium chloride.

Červa's formula of axenic medium is

Bacto-Casitone® (Difco)	20 g
Calf or horse serum	100 mℓ
Penicillin	500,000 units
Streptomycin	50,000 mcg
Bidistilled water	900 mℓ

D. Modification of Červa's Medium for *N. fowleri*

The original formula of Červa was modified by the late Dr. E. Willaert with excellent results.[66] This modification is especially good for axenic culture. The temperature should be 37°C with a pH of 7.

Bacto-Casitone® (Difco)	20 g
Folic acid	2 mg
Glucose	1 g
Fetal calf serum (Flow)	50 mℓ
Penicillin	500,000 units
Streptomycin	50,000 mcg
Bidistilled water	950 mℓ

In addition, the late E. Willaert also used the following medium with good results, mainly for the study of the antigenic structures of the genera *Naegleria* and *Acanthamoeba* spp.

Bacto-Casitone® (Difco)	20 g
Sodium chloride	5 g
Glucose	1 g
Folic acid	2 mg
Biotin	20 mcg
Penicillin	500,000 units
Streptomycine	50,000 mcg
Bidistilled water	1000 mℓ
pH	7

This medium should be filtered and sterilized. The incubation growth temperature should be 37°C and may be stored in the refrigerator at 4°C.

E. Chang Medium for Free-Living Amebas

According to Chang, small, free-living amebas grow well on buffered sucrose tryptose (BST) agar in association with Gram-negative bacteria.[15-18] Dr. Chang used this and other media, like broths with a heterogenous group of bacteria with success for culture, cloning, isolation, and maintenance of free-living amebas.

V. INOCULATION IN THE EXPERIMENTAL ANIMAL

The biological specimens may also be intracerebrally, intraperitoneally, or intranasally inoculated in experimental animals.[7] The mouse is the best animal model. If the laboratory animals develop the clinical symptoms of amebic infection (cerebral, pulmonary, or corneal), cultures and histological examination of the tissues should be made to identify the trophozoites or cysts and to assess the degree of extension of pathological changes. The procedure for animal inoculation and the results are fully explained in Chapter 9.

VI. TISSUE CULTURE

Pathogenic free-living amebas can be placed into mammalian cell cultures (Rhesus monkey kidney cells, chicken fibroblasts, vero cells, He LA cells, mouse embryo cells, mouse and human fibroblasts, and tissue culture of different tumors.[33,39-41,69,70] The amebas may grow very well and may produce cytopathic effects destroying the cells in the culture.[13,14,19] This is an excellent test to prove the pathogenic or virulent potential of such amebas. Brown has demonstrated direct phagocytic activity of the amebic trophozoites against the cells in culture.[4]

Originally, a transmissible agent virus named "lipovirus"[61] and Ryan virus[2,51] was suspected to cause the cytopathic effect in cell cultures. However, later it was demonstrated that such "lipovirus or toxic transmissible agents" were, in reality, free-living amebas that accidentally contaminated the cell culture.[13,14]

Naegleria spp. produce an "enzyme" or a "toxic factor" which is present in cell-free lysates. These lysates can provoke degenerative changes in the cell culture. This factor present in the cell-free lysate can be serially passed and plays an important role in the pathogenesis of PAM caused by *N. fowleri.*[32]

Legionella pneumophila has been found to be associated with free-living *Acantham-*

oeba. This suggests the possibility that free-living amebas may be carriers of *L. pneumophila.*[1,23,53,61] Tissue cultures of neoplasms (human or animal) have been reported as contaminated with free-living amebas.[2,13,14,51,69,70] In other instances, it has been demonstrated that cytopathic effects were produced by pathogenic *N. fowleri* or *N. gruberi* under appropriate conditions.[24]

VII. SEROLOGICAL AND IMMUNOLOGICAL METHODS

Different types of antibodies are involved in the antiamebic humoral immune response. The immunoglobulins that probably prevent the spread of infection are IgA, IgG, and IgM. Cell-mediated immunity may play an important role in the prevention of amebic disease and protection of the host.[27] Exposure to nonpathogenic species of free-living amebas can produce antibodies that may protect the host against infection.[6] Sera from young adult humans contains specific agglutinating activity for *N. fowleri,* but the agglutinating activity for infants is negligible. Apparently, there is no transplacental passage of the antibodies.[52] Host immunity probably exists and may be developed in humans and animals.

So far, no reliable diagnostic test has been found to diagnose PAM or GAE by serological methods. However, agglutination of *N. fowleri* and *N. gruberi* by antibodies present in human serum has been reported;[10,27] apparently there is no cross reactivcross reactivity between them.[52] Starch gel electrophoresis is an effective method for identification of pathogenic and nonpathogenic *Naegleria* spp.[47] Isoenzyme patterns of hydrosoluble proteins may be used to classify free-living amebas and to differentiate the pathogenic from the nonpathogenic.[28,29,36,63] In addition, the difference in Concanavalin A agglutination and the specific activity of acid phosphatase and leucine amino peptidase are reliable and simple methods for separation of *N. fowleri* from nonpathogenic *Naegleria* spp. in axenic cultures.[30] Isoenzyme and total protein analysis by agarose isoelectric focusing can be used also for classifying *Acanthamoeba* spp.[31]

Complement fixation tests and the direct hemagluttination test have been devised for the detection of specific antibodies against pathogenic free-living amebas of the genera *Naegleria* and *Acanthamoeba* spp.[10]

Kenney[38] demonstrated increased CSF antibody titers (complement fixation antibodies) against *Acanthamoeba* spp. in two patients. This was the first indication that antibodies may be produced in individuals when they harbor the protozoa or have suffered the disease.

Ferrante and Rowan-Kelly have demonstrated that the normal human serum contained an amebicidal property for *A. culbertsoni.*[33] Visvesvara and collaborators also demonstrated the presence of anti-*Acanthamoeba* antibodies in human serum. They collected 150 sera from the staff of an elementary school, and by immunofluorescence antibody techniques demonstrated that 13% of the collected samples contained anti-*Acanthamoeba* spp. antibodies.[65]

Cursons and collaborators demonstrated, in a series of well-prepared experiments, the role of humoral antibodies and cell-mediated immunity in the protection against free-living amebic infection.[27]

Precipitin antibody titers have also been demonstrated in other cases suffering *A. polyphaga* keratitis.[62] Hypersensitivity pneumonitis has been demonstrated in experimental animals.[60] More details of immunoelectrophoretical analysis (EPA) and immunoabsorption (IMA) of hydrosoluble proteins of free-living amebic trophozoites are given in Chapter 8.

REFERENCES

1. Anand, C. M., Skinner, A. R., Malic, A., and Kurtz, J. B., Interaction of *L. pneumophila* and a free-living ameba *(Acanthamoeba palestinensis)*, *J. Hyg.*, 91(2), 167, 1983.

2. Armstrong, J. A. and Pereira, M. S., Identification of "Ryan virus" as an ameba of the genus *Hartmannella*, *Br. Med. J.*, 1, 212, 1976.

3. Band, R. N., Nutritional and related biological studies on the free-living soil amoeba. *Hartmannella rhysodes*, *J. Gen. Microbiol.*, 21, 80, 1959.

4. Brown, T., Observations by light microscopy on the cytopathogenicity of *Naegleria fowleri* in mouse embryo-cell cultures, *J. Med. Microbiol.*, 11, 249, 1978.

5. Cailleau, R., Utilization des mileux liquides par *Acanthamoeba castellanii*, *C. R. Soc. Biol. (Paris)*, 116, 721, 1934.

6. Červa, L., Immunological studies on *Hartmannellid* amoebae, *Folia Parasitol. (Praha)*, 14, 19, 1967.

7. Červa, L., Intracerebral inoculation of experimental animals in pathogenetical studies of *Hartmannella castellanii*, *Folia Parasitol. (Praha)*, 14, 171, 1967.

8. Červa, L., Amoebic meningoencephalitis: axenic culture of *Naegleria*, *Science*, 163, 576, 1969.

9. Červa, L., The influence of temperature on the growth of *Naegleria fowleri* and *Naegleria gruberi* in axenic culture, *Folia Parasitol. (Praha)*, 24, 221, 1977.

10. Červa, L., Detection of antibody against *Limax* amoebae by means of the indirect haemagglutination test, *Folia Parasitol. (Praha)*, 24, 293, 1977.

11. Červa, L., Some further characteristics of the growth of *Naegleria fowleri* and *N. gruberi* in axenic culture, *Folia Parasitol. (Praha)*, 25, 1, 1978.

12. Červa, L., Laboratory diagnosis of primary amoebic meningoencephalitis and methods for the detection of Limax amoebae in the environment, *Folia Parasitol. (Praha)*, 27, 1, 1980.

13. Chang, R. S., Properties of a transmissible agent capable of inducing marked DNA degradation and thymine catabolism in a human cell, *Proc. Soc. Exp. Biol. Med.*, 107, 135, 1961.

14. Chang, R. S., Pan, I., and Rosenau, B. J., On the nature of the "lipovirus", *J. Exp. Med.*, 124, 1153, 1966.

15. Chang, S. L., Cultural, cytological and ecological observations on the amoeba stage of *Naegleria gruberi*, *J. Gen. Microbiol.*, 18, 565, 1958.

16. Chang, S. L., Cytological and ecological observations on the flagellate transformation of *Naegleria gruberi*, *J. Gen. Microbiol.*, 18, 579, 1958.

17. Chang, S. L., Growth of small free-living amoebae in various bacterial and in bacteria-free cultures, *Can. J. Microbiol.*, 6, 397, 1960.

18. Chang, S. L. Small, free-living amebas: cultivation, quantitation, identification, classification, pathogenesis, and resistance, *Curr. Top. Comp. Pathobiol.*, 1, 201, 1971.

19. Chi, L., Vogel, J., and Shelokov, A., Selective phagocytosis of nucleated erythrocytes by cytotoxic amoeba in cell culture, *Science*, 130, 1763, 1959.

20. Cleland, P. G., Lawande, R.V., Onyemelukwe, G., and Whittle, H. C., Chronic amebic meningoencephalitis, *Arch. Neurol.*, 39, 56, 1982.

21. Cline, M., Marciano-Cabral, F., and Bradley, S. G., Comparison of *Naegleria fowleri* and *Naegleria gruberi* cultivated in the same nutrient medium, *J. Protozool.*, 30, 387, 1983.

22. Culbertson, C. G., Ensminger, P. W., and Overton, W. M., The isolation of additional strains of pathogenic *Hartmannella* sp. *(Acanthamoeba)*. Proposed culture method for application to biological material, *Am. J. Clin. Pathol.*, 43, 383, 1965.

23. Culbertson, C. G. and Harper, K., Surface coagglutination with formalinized stained protein A. staphyloccoci in the immunologic study of three pathogenic amebae, *Am. J. Trop. Med. Hyg.* 29, 785, 1980.

24. Cursons, R. T. M. and Brown, T. J., Use of cell cultures as an indicator of pathogenicity of free-living amoebae, *J. Clin. Pathol.*, 31, 1, 1978.

25. Cursons, R. T. M., Brown, T. J., and Keys, E. A., Diagnosis and identification of the aetiological agents of primary amoebic meningoencephalitis (PAM), *N. Z. J. Med. Lab. Technol.* 32, 11, 1978.

26. Cursons, R. T. M., Donald, J. J., Brown, T. J., and Keys, E. A., Cultivation of pathogenic and nonpathogenic free-living amebae, *J. Parasitol.*, 65, 189, 1979.

27. Cursons, R. T. M., Brown, T. J., Keys, E. A., Moriarty, K. M., and Till, D., Immunity to pathogenic free-living-amoebae. Role of humoral antibody, *Infect. Immun.*, 29, 401, 1980.

28. Daggett, P. M. and Nerad, T. A., The biochemical identification of *Vahlkampfiid* amoebae, *J. Protozool.*, 30, 126, 1983.

29. De Jonckheere, J. F., Isoenzyme patterns of pathogenic and nonpathogenic *Naegleria* spp. using agarose isoelectric focusing, *Ann. Microbiol. (Paris)*, 133, 319, 1982.

30. De Jonckheere, J. F. and Dierickx, P. J., Determination of acid phosphatase and leucine amino peptidase activity as an identification method for pathogenic *Naegleria fowleri*, *Trans. R. Soc. Trop. Med. Hyg.*, 76, 773, 1982.

31. De Jonckheere, J. F., Isoenzyme and total protein analysis by agarose isoelectric focusing and taxonomy of the genus *Acanthamoeba*, *J. Protozool.*, 30, 701, 1983.

32. Dunnebacke, T. H. and Schuster, F. L., The nature of a cytopathogenic material present in amebae of the genus *Naegleria*, *Am. J. Trop. Med. Hyg.*, 26, 412, 1977.

33. Ferrante, A. and Rowan-Kelly, B. C., Activation of the alternative pathway of complement by *Acanthamoeba culbertsoni*, *Clin. Exp. Immunol.*, 54, 477, 1983.

34. Griffin, J. L., Temperature tolerance of pathogenic and nonpathogenic free-living amoebas, *Science*, 178, 869, 1972.

35. Griffin, J. L., The pathogenic amoeboflagellate *Naegleria fowleri:* environmental isolations, competitors, ecologic interactions and the flagellate-empty habitat hypothesis, *J. Protozool.*, 30, 403, 1983.

36. Haight, J. B. and John, D. T., Varying the serum component in axenic cultures of *Naegleria fowleri*, *Proc. Helminthol. Soc. Wash.*, 49, 127, 1982.

37. Hoffler, A. S. and Rubel, L. R., Free-living amoebae identified by cytologic examination of gastorintestinal washings, *Acta Cytol.*, 18, 59, 1974.

38. Kenney, M., The micro-Kolmer complement fixation test in routine screening for soil ameba infection, *Health Lab. Sci.*, 8, 5, 1971.

39. Marciano-Cabral, F. M. and Bradley, S. G., Cytopathogenicity of *Naegleria gruberi* for rat neuroblastoma cell cultures, *Infect. Immun.*, 35, 1139, 1982.

40. Marciano-Cabral, F. M., Patterson, M., John, D. T., and Bradley, S. B., Cytopathogenicity of *Naegleria fowleri* and *Naegleria gruberi* for established mammalian cell cultures, *J. Parasitol.*, 68, 1110, 1982.

41. Marciano-Cabral, F. and John, D. T., Cytopathogenicity of *Naegleria fowleri* for rat neuroblastoma cell cultures: scanning electron microscopy study, *Infect. Immun.*, 40, 1214, 1983.

42. Martinez, A. J., dos Santos, J. G., Nelson, E. C., Stamm, W. P., and Willaert, E., Primary amebic meningoencephalitis, in *Pathology Annual*, Vol. 12, Sommers, S. C. and Rosen, P. P., Eds., Appleton-Century-Crofts, New York, 1977, 225.

43. McCool, J. A., Spudis, E. V., McLean, W., White, J., and Visvesvara, G. S., Primary amebic meningoencephalitis diagnosed in the emergency department, *Ann. Emergency Med.*, 12, 35, 1983.

44. Medley, S., Acridine orange: method for diagnosis of amebic meningitis (correspondence), *Med. J. Aust.*, 2, 635, 1980.

45. Napolitano, J. J. and Gamble, H. R., Folic acid stimulation of axenically grown, *Naegleria fowleri*, *Protistologica*, 14, 183, 1978.

46. Neff, R. J., Purification, axenic cultivation and description of a soil amoeba, *Acanthamoeba* sp., *J. Protozool.*, 4, 176, 1957.

47. Nerad, T. A. and Daggett, P. -M., Starch gel electrophoresis: an effective method for separation of pathogenic and nonpathogenic *Naegleria* strains, *J. Protozool.*, 26, 613, 1979.

48. Nerad, T. A., Visvesvara, G., and Daggett, P. M., Chemically defined media for the cultivation of *Naegleria*. Pathogenic and high temperature tolerant species, *J. Protozool.*, 30, 383, 1983.

49. O'Dell, W. and Stevens, A. R., Quantitative growth of *Naegleria* in axenic culture, *Appl. Microbiol.*, 25, 621, 1973.

50. Page, F. C., Taxonomic criteria for *limax* amoebae with descriptions of 3 new species of *Hartmannella* and 3 of *Vahlkampfia*, *J. Protozool.*, 14, 449, 1967.

51. Pereira, M. S., Marsden, H. B., Corbitt, G., and Tobin, J. O., Ryan virus: a possible new human pathogen, *Br. Med. J.*, 1, 130, 1966.

52. Reilly, M. F., Marciano-Cabral, F., Bradley, D. W., and Bradley, S. G., Agglutination of *Naegleria fowleri* and *Naegleria gruberi* by antibodies in human serum, *J. Clin. Microbiol.*, 17, 576, 1983.

53. Rowbotham, T. J., Preliminary report on the pathogenicity of *Legionella pneumophila* for fresh water and soil amoeba, *J. Clin. Pathol.*, 33, 1179, 1980.

54. Saygi, G. and Warhurst, D. C., Greensteins five dye stain. A rapid and simple differential stain for amoebae, *Trans. R. Soc. Trop. Med. Hyg.*, 64, 19, 1970.

55. Schuster, F. L. and Dunnebacke, T. H., Growth at 37°C of the EG, strain of the amebo-flagellate *Naegleria gruberi* containing virus-like particles. I. Nuclear changes, *J. Invertebrate Pathol.*, 23, 172, 1974.

56. Schuster, F. L. and Dunnebacke, T. H., Growth at 37°C of the EG, strain of the amebo-flagellate *Naegleria gruberi* containing virus-like particles. II. Cytoplasmic changes, *J. Invertebrate Pathol.*, 23, 182, 1974.

57. Singh, B. N., Nuclear division in nine species of small free-living amoebae and its bearing on the classification of the order *Amoebida*, *Philos. Trans. R. Soc. London Ser. B*, 236, 405, 1952.

58. Stamm, W. P., The staining of free-living amoeba by indirect immunofluorescence, *Ann. Soc. Belge Med. Trop.*, 54, 321, 1974.

59. Stevens, A. R., DeJonckheere, J., and Willaert, E., *Naegleria Lovaniensis* new species: isolation and identification of six thermophilic strains of a new species found in association with *Naegleria fowleri*, *J. Parasitol.*, 10, 51, 1980.

60. Sykora, J., Karol, M., Kelete, G., and Novak, D., Amoebae as sources of hypersensitivity pneumonitis, *Environ. Int.,* 8, 343, 1982.

61. Tyndall, R. L. and Domingue, E. L., Cocultivation of *Legionella pneumophila* and free-living amoebae, *Appl. Environ. Microbiol.,* 44, 954, 1982.

62. Visvesvara, G. S., Jones, D. B., and Robinson, N. M., Isolation, identification, and biological characterization of *Acanthamoeba polyphaga* from a human eye, *Am. J. Trop. Med. Hyg.,* 24, 784, 1975.

63. Visvesvara, G. S., Free-living pathogenic amoebae, in *Manual of Clinical Microbiology,* 3rd ed., Lennette, E. H., Balows, A., Hausler, W. J., and Truant, J. P., Eds., American Society of Microbiology, Washington, D.C., 1980.

64. Visvesvara, G. S., Mirra, S. S., Brandt, F. H., Moss, D. M., Mathews, H. M., and Martinez, A. J., Isolation of two strains of *Acanthamoeba castellanii* from human tissue and their pathogenicity and isoenzyme profiles, *J. Clin. Microbiol.,* 18, 1405, 1983.

65. Visvesvara, G. S., Brandt, F. H., Baxter, P. J., and Healy, G. R., Isolation of a pathogenic *Acanthamoeba polyphaga* from disposable filters attached to heating, ventilation, and air conditioning (HVAC) units and demonstration of anti-*Acanthamoeba* antibody in human sera, *J. Protozool.,* 29 (Abstr.), 489, 1982.

66. Willaert, E., Isolement et culture in vitro des amibes du genre *Naegleria, Ann. Soc. Belge Med. Trop.,* 51, 701, 971.

67. Willaret, E., Jadin, J. B., and LeRay, D., Structures immunochimiques comparées d'amibes du genre *Naegleria, Protistologica,* 8, 497, 1972.

68. Willaert, E. and Stevens, A. R., Indirect immunofluorescent identification of *Acanthamoeba* causing meningoencephalitis, Pathol. Biol. (Paris), 24, 545, 1976.

69. Willaert, E., Stevens, A. R., and Tyndall, R. L., *Acanthamoeba royreba* sp.n. from a human tumor cell culture, *J. Protozool.,* 25, 1, 1978.

70. Willaert, E., Stevens, A. R., and Tyndall, R. L., Identification of *Acanthamoeba culbertsoni* from cultured tumor cells, *Protistologica,* 14, 319, 1978.

Chapter 8

RETROSPECTIVE DIAGNOSIS: PAM AND GAE

I. INTRODUCTION

Retrospective diagnoses of free-living amebic infections have been made while examining sections from CNS tissues fixed in formalin, embedded in paraffin, and stained with hematoxylin and eosin (H & E).[2,10,16,23,24] In such instances, identification of the responsible amebas has rested chiefly on tenuous morphological features such as size and shape of the trophozoites, and on the shapes, size, and other structural details of the cysts.[11]

The specific immunological characteristics of free-living amebas open up new and exciting perspectives for taxonomical, epidemiological, morphological, and clinical investigations.[18,19] This is particularly true when one considers a complex group of protozoa that cannot be well differentiated morphologically.

With the indirect immunofluorescent antibody technique (IFAT), with the immunoperoxidase method (IPA), with the highly specific immune sera, the immunoelectrophoresis (IEPA), and the enzyme-linked immunoabsorbent assay (ELISA), the nature, taxonomical position, and composition of a wild protozoan population might be determined rapidly and effectively. However, there is considerable antigenic overlap between species of the genus *Acanthamoeba* and *Naegleria,* but, in spite of that, antisera may be produced in rabbits with highly specific selectivity by suitable immunoabsorption technique.[5] In addition, retrospective diagnosis may be accomplished quite accurately with the determination of the etiologic amebic agent.

II. IMMUNOFLUORESCENT ANTIBODY TECHNIQUE (IFAT)

The IFAT is based on the fact that the antibody molecule can be conjugated with fluorescent compounds, producing a brilliant, fluorescent protein that precipitates over the areas where the antigen is located when observed under a fluorescence microscope.[4,9,12,14,22]

The IFAT of CNS sections is a valuable procedure for the identification and localization of free-living amebas in patients who have died of suspected amebic meningoencephalitis[3] (Figures 1 and 2).

Specific antibodies for free-living species can be elicited in rabbits, and a rapid diagnosis in human and animal CNS or other tissues can be accomplished.

The histologic material should be formalin-fixed to ensure the best results. If tissues are fixed in mercury fixatives, they cannot be stained by the IFAT unless the mercury is first removed with alcohol-iodine; even then the fluorescence is weaker than if the same tissue had been fixed in formalin. The common histologic sections (5 μm thick) are used to perform the IFAT after deparaffinization of the tissues.

In addition to its value in clinical diagnosis, IFAT provides a rapid screening method for the detection of pathogenic free-living amebas in swimming pools, tap water, and other domestic and recreational water supplies.

In the indirect immunofluorescent method,[13,15] the antigen is incubated with different dilutions of the antisera. When the antisera contains specific antibodies to the antigen, these will be fixed on the antigenic sites. The immunocomplex is visualized with fluorescein-conjugated antiimmunoglobulins. This technique benefits from a greater sensitivity than the direct method. The results are read with a Leitz Ortholux® microscope equipped for fluorescence with a Ploemopak 2® and a super pressure mer-

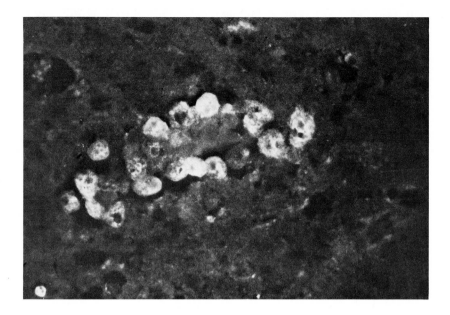

FIGURE 1.   Indirect immunofluorescent staining of CNS sections with *Naegleria fowleri* antiserum (dilution 1:256; performed by the late Dr. Eddy Willaert). (Magnification × 600.)

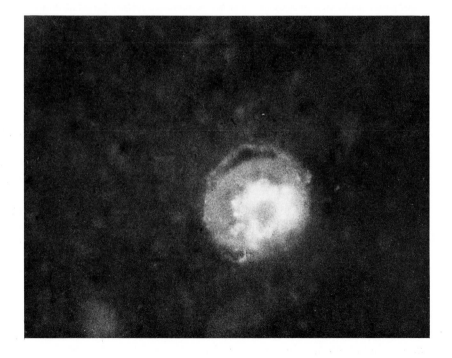

FIGURE 2.   Indirect immunofluorescent staining of a *Naegleria fowleri* trophozoite from a CNS section in a case of PAM. *N. fowleri* antiserum (dilution 1:32; performed by the late Dr. Eddy Willaert). (Magnification × 800.)

cury lamp HBO 50W. The filter system used is for FITC combined with two interference filters KP 490, K 510, K 530, and K 445.

III. IMMUNOPEROXIDASE METHOD (IPA)

The IPA is based on the use of the enzyme horseradish peroxidase as a marker to visualize antigens at the cellular surface level. The histochemical reaction between peroxidase and its substrate gives an insoluble reaction visible by light and electron microscopy.[6-8] Culbertson, in 1975, used this method for the demonstration of both *Naegleria fowleri* and *Acanthamoeba* spp. in the tissues of patients who died with suspected free-living amebic disease[6] (Figure 3). The method can also be used in experimental animals (Figure 4).

The IPA method is easy to standardize, the preparation of peroxidase conjugates is not complicated, permanent preparations can be made, and a conventional light microscope can be used. However, these procedures are not yet frequently used for the diagnosis of free-living amebas. The IPA method by light and electron microscopy may be used to demonstrate specific plasma membrane antigens from *A. castellanii, A. culbertsoni,* or *Naegleria* spp. In the future this method may be shown to be more valuable and reliable than the IFAT, particularly the one that will use the specific monoclonal antibodies.[18]

IV. IMMUNOELECTROPHORESIS (IEPA)

Immunoelectrophoretical analysis (IEPA) of hydrosoluble protein of free-living amebas belonging to the genera *Naegleria* and *Acanthamoeba* allows identification and separation of the different species belonging to these genera.[19-21] This analysis is an excellent procedure to clarify and determine the relationship and antigenic specifications of a free-living ameba. The IEPA allows the characterization of soluble and immunogenic cellular components, and antigenic patterns. Two techniques are involved in this immunochemical analysis:

1. IEPA and gel diffusion which allows the characterization of the antigenic components in the mixture
2. Immunoabsorption (IMA) which allows determination of the relationship and specifications of the components

The IEPA has two theoretical limits: first, it is only accessible for soluble antigens which represent a significant part of an amebic organism, and, second, the method is essentially qualitative allowing only the establishment of absolute differences of antigenic specificities. Two-dimensional electrophoresis can be employed to palliate this inconvenience. Further details of the immunoelectrophoretical analysis and its variants are given in Chapters 3 and 7.

V. PREPARATION OF THE ANTISERUM

Two methods may be used to prepare the antisera against *Acanthamoeba* and *Naegleria* species. They may be prepared either by multiple intradermal injections of soluble or insoluble extracts or cell surface antigens of ameba or by intravenous inoculation of whole living amebas. The rabbits are bled a short time after immunization. The antisera obtained after intradermal immunization can reach a high titer, but then shows important cross reactions, while after the intravenous route, the titer of the antisera is low but less cross reaction is observed.[17]

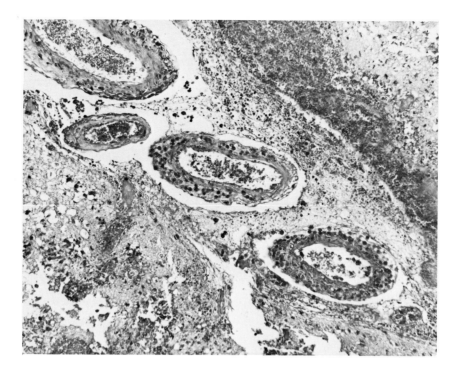

FIGURE 3.   Immunoperoxidase staining of CNS section in a case of GAE. Numerous trophozoites and cysts can be identified within the vascular walls. The *Acanthamoeba castellanii*
antiserum at dilution 1:20 = 4+; 1:50 = 2+, and 1:100 = 1+. The IPA reaction was negative
for anti-*A. culbertsoni,* anti-*A. polyphaga,* and anti-*A. astronyxis* (performed by Dr. C. G.
Culbertson). (Magnification × 100.)

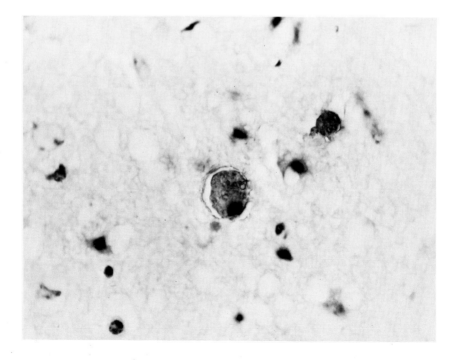

FIGURE 4.   Immunoperoxidase stain of CNS tissue of a mouse 10 days postintranasal inoculation with pathogenic *N. australiensis* demonstrating positive staining of amebas (dilution
1:320; performed by J. de Jonckheere). (Magnification × 1000.)

Axenic free-living amebic cultures, age of about 5 days, are washed several times in phosphate buffered saline (NaCl, 8.5 g; $Na_2 HPO_4$, 3.5 g; bidistilled water, 1000 ml; pH 7.0) and centrifuged for 5 min at 885 g. The amebas are then fixed for 30 min in a 2% formalin solution made in buffered saline. After fixation the cells are washed again three times. A dilution is made to obtain about 25 ameba per microscopic field (obj. 40 × sc. 10×). Drops of this diluted material are placed on special slides for immunofluorescence allowing ten dilutions per slide. After drying the antigen, the slides can be stored at 25°C for several months.

With the hyperimmune antisera produced in rabbits and the hydrosoluble extracts of several species of *Naegleria* and *Acanthamoeba* spp. and *Vahlkampfia*, it will be possible to establish some epidemiological features and taxonomic criteria concerning PAM and GAE. The classification of free-living amebas is essentially based on morphological characteristics. With the aid of IEPA, the importance of antigenic relationship and specificities of the free-living amebas belonging to the genera *Naegleria* and *Acanthamoeba* at the level of the genus, the family, and phylum may be evaluated and determined. However, this analysis is time consuming and cannot be considered a routine laboratory technique since it is neither easy nor rapid to perform. For this reason, a rapid identification of free-living amebas by simple and sensitive immunological techniques, such as IFAT and IPA staining, is more suitable.[6-8,21]

REFERENCES

1. Avrameas, S., Indirect immunoenzyme technique for the intracellular detection of antigens, *Immunochemistry*, 6, 825, 1969.
2. Callicott, J. H., Amebic meningoencephalitis due to free-living amebas of the *Hartmannella (Acanthamoeba) Naegleria* group, *Am. J. Clin. Pathol.*, 49, 84, 1968.
3. Carter, R. F., Cullity, G. J., Ojeda, V. J., Silverstein, P., and Willaert, E., A fatal case of meningoencephalitis due to a free-living amoeba of uncertain identity. Probably *Acanthamoeba* sp., *Pathology*, 13, 51, 1981.
4. Červa, L., Use of fluorescent antibody technique to identify pathogenic *Hartmannellae* in tissues of experimental animals, *Folia Parasitol. (Praha)*, 13, 328, 1966.
5. Červa, L. and Kramar, J., Antigenic relationships among several *limax* amoebae isolates assessed with the indirect fluorescent antibody test (IFAT), *Folia Parasitol. (Praha)*, 20, 113, 1973.
6. Culbertson, C. G., Soil ameba infection. Specific indirect immunoenzymatic (peroxidase) staining of formalin-fixed paraffin sections, *Am. J. Clin. Pathol.*, 63, 475, 1975.
7. Culbertson, C. G., Immunoperoxidase staining of *E. histolytica* and soil amebas in formalin-fixed tissue, *Am. J. Clin. Pathol.*, 67, 529, 1977.
8. Cursons, R. T. M., Brown, T. J., and Culbertson, C. G., Immunoperoxidase staining of trophozoites in primary amoebic meningoencephalitis, *Lancet*, ii, 479, 1976.
9. De Jonckheere, J., De Van Dijck, P., and Van de Voorde, H., Evaluation of the indirect fluorescent-antibody technique for identification of *Naegleria* species, *Appl. Microbiol.*, 28, 159, 1974.
10. Dos Santos, J. G., Fatal primary amebic meningoencephalitis: a retrospective study in Richmond, Virginia, *Am. J. Clin. Pathol.*, 54, 737, 1970.
11. Miller, G., Cullity, G., Walpole, I., O'Connor, J., and Masters, P., Primary amoebic meningoencephalitis in Western Australia, *Med. J. Aust.*, 1, 352, 1982.
12. Nakane, P. K. and Pierce, G. B., Enzyme-labeled antibodies: preparation and application for the localization of antigens, *J. Histochem. Cytochem.*, 14, 929, 1966.
13. Parelkar, S. A., Stamm, W. P., and Hill, K. R., Indirect immunofluorescent staining of *Entamoeba histolytica* in tissues, *Lancet*, 1, 212, 1971.
14. Siddiqui, W. A. and Balamuth, W., Serological comparison of selected parasitic and free-living amoebae in vitro using diffusion-precipitation and fluorescent-antibody technics, *J. Protozool.*, 13, 175, 1965.
15. Stamm, W. P., The staining of free-living amoebae by indirect immunofluorescence, *Ann. Soc. Belge Med. Trop.*, 54, 321, 1974.

16. Symmer, W. St. C., Primary amoebic meningoencephalitis in Britain, *Br. Med. J.,* 4, 449, 1969.
17. Vaitukaitis, J., Robbins, J. B., Nieschlag, E., and Ross, G. T., A method for producing specific antisera with small doses of immunogen, *J. Clin. Endocrinol.,* 33, 988, 1971.
18. Visvesvara, G. S., Peralta, M. J., and Brandt, F. H., Development of species specific monoclonal antibodies to *Naegleria fowleri* (abstr.), *J. Protozool.,* 30, 14A, 1983.
19. Willaert, E., Jadin, J. B., and LeRay, D., Comparative antigenic analysis of *Naegleria* species, *Ann. Soc. Belge Med. Trop.,* 53, 59, 1973.
20. Willaert, E., Jamieson, A., Jadin, J. B., and Anderson, K., Epidemiological and immunoelectrophoretic studies on human and environmental strains of *Naegleria fowleri, Ann. Soc. Belge Med. Trop.,* 54, 333, 1974.
21. Willaert, E., Étude immunotoxonomique des genres *Naegleria* et *Acanthamoeba (Protozoa: Amoebida).* Thèse Doctorat en Sciences. Université de Lille. *Acta Zool. Pathol. Antverp.* 65, 1976.
22. Willaert, E. and Stevens, A. R., Indirect immunofluorescent identification of *Acanthamoeba* causing meningoencephalitis, *Pathol. Biol.,* 24, 545, 1976.
23. Willaert, E., Stevens, A. R., and Healy, G. R., Retrospective identification of *Acanthamoeba culbertsoni* in a case of amebic meningoencephalitis, *J. Clin. Pathol.,* 31, 717, 1978.
24. Willaert, E., Stevens, A. R., and Healy, G. R., Retrospective identification of *Acanthamoeba culbertsoni* in a case of amebic meningoencephalitis, *J. Clin. Pathol.,* 31, 717, 1978.

Chapter 9

ANIMAL MODELS: PAM AND GAE

I. THE EXPERIMENTAL DISEASE

The relevance and usefulness of any animal model resides in its similarities to the human disease. Some or all of the histopathological features of the animal model of PAM and GAE show similarities with the human disease (Table 1). Because the mouse is probably the best animal model of PAM and GAE, numerous reports have been published based on this model to test the epidemiological, immunological, morphological, pathological, and therapeutic data.[1,3,8,9,12-15,17-21,23,27-30,32,33,35-37,40,55,59-64,66-69,70-76]

A. Naegleria sp. Infection (PAM)

The characteristic features of PAM due to Naegleria spp. (N. fowleri and probably others) in man have also been noted in experimental infection in animals — chiefly in the mouse — with the same portal of entry, incubation period, migration to the CNS, and identical histopathological features.[6,34,74] Germ-free guinea pigs[58] and other animals, some of them under immunosuppressive therapy, have been tested as animal models.[46,49,65,77,78] In addition, tissue and cell cultures have been used to test the virulence of different strains of amphizoic or free-living amebas.[4,5,7,10,24,25,31,38,39] Different routes of inoculation have also been tested from subcutaneous to intravenous and intranasal.

The use of the mouse as the animal model for experimental infection of free-living amebas provides an opportunity to examine the pathogenesis of the disease, the mechanism and probable migration of the protozoa to the CNS, the host immune mechanisms, the epidemiology and control of the disease, the histopathological changes within CNS and other organs affected (mainly lungs), and the effects of chemotherapeutic agents on the management and presentation of the infection.[32,75] This animal model should contribute greatly to a better understanding of free-living amebic infection.

Using pathogenic amebas isolated from the environment or from fatal human cases, a similar if not identical infection can be regularly induced in white mice by intranasal instillation of variable numbers of viable trophozoites. With these inocula, some mice develop the disease and die soon afterwards.

Experiments with mice have shown that some animals acquired and died of PAM or GAE even when exposed intranasally to only a few trophozoites. However, it is true that the higher the number of pathogenic N. fowleri used in the inoculum, the faster the development of the disease.

In humans we do not know if only one ameba or a few amebas are necessary to cause the infection. The probability of infection may depend on the number or concentration of amebas in the inoculum and the duration of exposure.

In addition to CNS infections, some strains of N. fowleri and other Naegleria species of low pathogenicity may produce acute and subacute pneumonitis followed by PAM.[26] Some strains of N. fowleri may decrease their virulence and may be capable of causing subacute or chronic encephalitis by long-term maintenance in axenic culture.[27] So far, all the amebas isolated from fatal human cases of PAM have been found to be ameboflagellates of the genus Naegleria.

The use of the mouse as an experimental animal model for studying free-living amebic infection is appropriate since the existence of this disease in man was first suggested following the discovery of fatal meningoencephalitis in mice after intranasal inocula-

Table 1
NAEGLERIA AND *ACANTHAMOEBA* SP. INFECTIONS: COMPARISON AND CONTRAST

	Naegleria	*Acanthamoeba*
Incubation period (days)	3—5	Probably >10
Clinical course (days)	Acute, fulminant (<7)	Subacute (8—30) Chronic (>32)
Portal of entry	Olfactory neuroepithelium	Skin, lung? Olfactory neuroepithelium
CNS spread	Direct, olfactory amyelinic nervous plexus	Probably hematogenous
Organs affected	Olfactory neuroepithelium and brain	Brain, skin, eyes, lungs
Host response	Acute necrotizing, hemorrhagic meningoencephalitis	Necrotizing "granulomatous" chronic encephalitis with multinucleated giant cells
CNS amebic forms	Trophozoites	Trophozoites and cysts

tion of a species of *Acanthamoeba*. When PAM was subsequently reported in humans, the remarkable similarity of the disease occurring in man to that produced experimentally was promptly recognized.[11,16,22] The mouse possesses a well-developed olfactory system with unique anatomical features and turbinates similar to humans. The close proximity of the olfactory neuroepithelium to CNS structures is well demonstrated in Figures 1, 2, and 3. The turbinates are lined by respiratory epithelium. The cribiform plate is pierced by nonmyelinated olfactory nerves, where the pathogenic ameba multiply and invade passing to the subarachnoid space (Figures 1 to 3). Figure 4A shows a head of a normal mouse in which the scalp and skull bones have been removed demonstrating the close relationship between the olfactory bulbs and the cerebral hemispheres and the cerebellum. Figure 4B shows another mouse which has been intranasally inoculated with pathogenic *N. fowleri* 5 days earlier. There is severe edema of CNS tissues associated with necrosis and hemorrhage (Figure 5).

The basic features of the disease in man have all been noted in experimental infections in the mouse. The animal model discloses the same incubation period and portal of entry, residence of amebas in the olfactory mucosa with loss of cilia and with invasion, and migration through submucosal structures and into amyelinic olfactory nerve plexuses. Then the amebic trophozoites pass through pores of the cribriform plate into the subarachnoid space with subsequent invasion of olfactory bulbs and frontal lobes with spread of the amebas to more distant areas of the brain (Figures 4B, 5, and 6). Numerous amebic organisms may be seen invading sustentacular cells, between sustentacular and sensory cells and other intercellular spaces, around small blood vessels, and within the submucosal unmyelinated nervous plexuses. At the surface of the nasal olfactory epithelium, swelling and partial disintegration of microvilli, sensory cilia, and kinocilia are frequently noted, especially in those cells in direct contact with the protozoa (Figure 6). Frequent aggregation of amebas in perivascular spaces and a scant neurotrophilic cellular response associated with and superimposed upon widespread areas of hemorrhagic necrosis are usually found.[51,52] The tissues infected with the pathogenic amebas can be processed for light or electron microscopic study.

Gray and white matter are both affected, and the pathologic changes are characterized by an acute inflammatory reaction associated with hemorrhage, edema, disintegration of neural structures, and widespread invasion by amebas. Trophozoites may be seen in the perivascular spaces adjacent to the adventitia of arterioles and capillaries

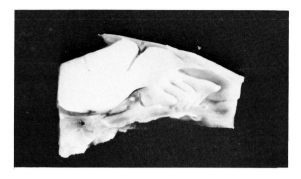

FIGURE 1.   Sagittal section of a mouse head demonstrating the nasal cavity (N) with the turbinates, the area of cribiform plate, and the brain (B). Specimen fixed in formalin and dehydrated to be used for scanning electron microscopy.

FIGURE 2.   Nasal cavity of a mouse, inoculated intranasally 48 hr previously with pathogenic *Naegleria fowleri*. There is denudation of the nasal mucosa (arrows). (Scanning EM; magnification × 100.)

or within the neuropil without inflammatory response (Figure 7). Mitotic activity of amebic trophozoites may be seen (Figure 8).

1. Potential Usefulness of the Animal Model

Important problems with the animal model of PAM to be studied and clarified are the mechanisms of penetration by amebas through the nasal and olfactory epithelium, immunological features, factors regulating proliferation of amebas within the nasal mucosa, host factors involved in invasion or spread of the protozoa, and effects of chemotherapeutic agents and antibiotics on preventing or controlling infection.[34]

N. fowleri loses virulence when maintained in prolonged in vitro culture. It has been

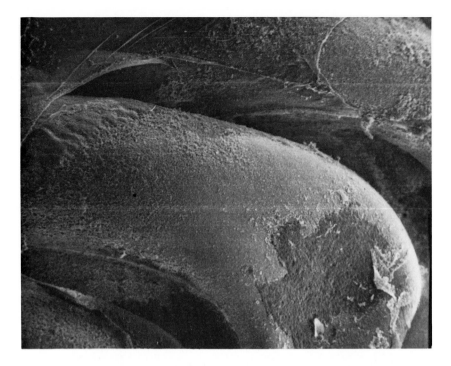

FIGURE 3.   One of the nasal turbinates demonstrating an area of mucosal denudation. (Scanning EM; magnification × 500.)

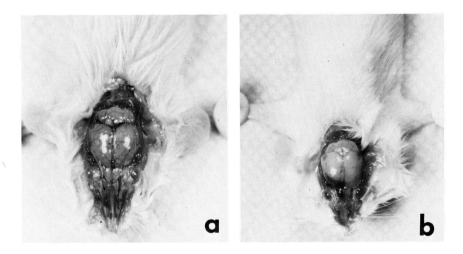

FIGURE 4.   (a) Head of a normal mouse. The scalp has been retracted and the upper portion of the skull has been removed for comparison. (b) Head of a mouse intranasally inoculated 5 days previously with 10,000 pathogenic *Naegleria fowleri.* The right olfactory bulb is hemorrhagic and the right cerebral hemisphere is edematous. (From Martinez, A. J., *Am. J. Pathol.,* 73(2), 545, 1973. With permission.)

suggested that this tendency may be attributed to nutritive deficiencies of the culture medium.[27,48] The loss of virulence can be reversed by a single passage to live mouse brain. Other animals may be susceptible to intranasal inoculation and may develop CNS involvement.

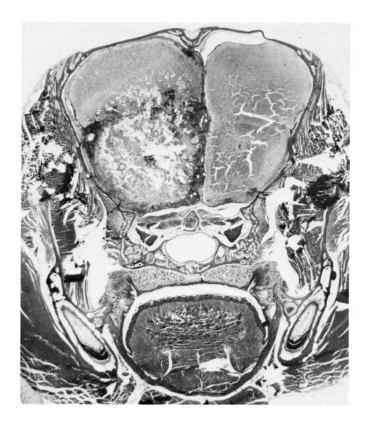

FIGURE 5.   Coronal section of a mouse head, 5 days after intranasal inoculation with pathogenic *Naegleria fowleri.* One cerebral hemisphere in severely edematous and necrotic. (H & E; magnification × 10.) (From Martinez, A. J., *Am. J. Pathol.,* 73(2), 545, 1973. With permission.)

B. *Acanthamoeba* spp. Infection (GAE)

How do the *Acanthamoeba* spp. reach the brain? After intranasal inoculation the mouse model may develop an Acanthamoebic pneumonitis and several weeks later the *Acanthamoeba* may reach the CNS, mainly the posterior fossa structures and basal ganglia sparing the olfactory bulbs (Figures 9 and 10). The amebic pneumonitis is prominent (Figures 11 and 12). Trophozoites and cysts of *Acanthamoeba* spp. may be found in the infected lung or brain (Figures 13A and B and 14A to D). Some amebic trophozoites may be encompassed by lymphocytes, histiocytes, and microglia. Phagocytic or microglial nodules may be found in the brain. Occasionally necrotic amebic trophozoites may be seen (Figure 14D). Cysts with the typical wrinkled double wall may be seen mainly in the lungs (Figure 15). Well-preserved amebic trophozoites surrounded by macrophages may be detected in the infected CNS tissue (Figure 16).

Experimental CNS and lower respiratory tract infections have been produced in the mouse model with both *N. fowleri* and *Acanthamoeba* spp. after treatment with spectrum antibiotics and corticosteroids. The lesions are mainly subacute and chronic with multinucleated giant cells.[56,57]

Sheep and other mammals have also been used in experimental infections.[65,77,78] The histopathological changes in the olfactory lobes are similar to those in the human disease. The findings indicated a hemorrhagic and necrotizing meningoencephalitis with necrosis in the olfactory lobes. Trophozoites may be present within CNS tissue in *Naegleria* spp. infections. Both trophozoites and cysts may be found within infected CNS tissues and lungs in *Acanthamoeba* spp. infection.

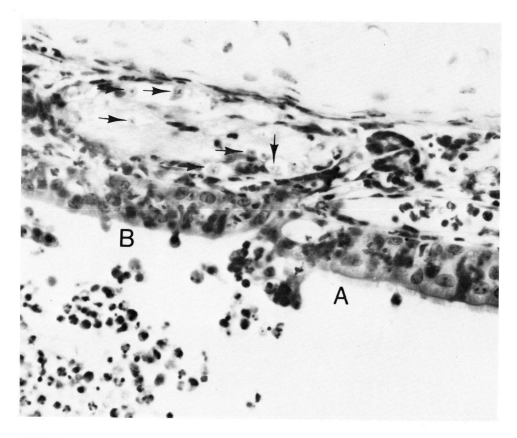

FIGURE 6.   Section at level of the cribiform plate of a mouse head 5 days after intranasal inoculation with 10,000 pathogenic *Naegleria fowleri*. Portion of the olfactory mucosa demonstrates the cilia (A). Other portion is devoid of cilia (B). Numerous amebic trophozoites can be seen within the amyelinic submucosal nervous plexus (arrows). (H & E; magnification × 400.) (From Martinez, A. J. et al., *Lab. Invest.*, 29(2), 121, 1973. With permission.)

C. Tasks for the Future and Research Ideas

Future challenges will be to use the nude mouse, immunosuppressed animals, inbred genetically controlled mice, and, perhaps, rats and other mammals to determine if they are more susceptible to the amebic infection.

The rationale for therapy and the study of the natural history of the disease may be tested in animal models and with them the factors that might induce virulence and loss of pathogenicity in free-living amebas.[27]

Lung lesions attributable to *N. fowleri* in animals inoculated with the organism intranasally have been well described.[46,49,75] However, it has been pointed out that the mouse lacked an amebicidal factor that is present in human serum, so it is unlikely that *N. fowleri* could directly invade human organs other than the CNS. It has been stated that *N. fowleri* is a neurotropic protozoon; but this may be relative because the CNS tissue possesses a very limited defense mechanism with a scant number of reticuloendothelial cells and microglia. Perhaps only a compromised host would allow involvement of organs such as the lungs and lymph nodes.

N. fowleri and *Acanthamoeba* spp. may be capable of causing subtle respiratory infections in man.[32,75] From the lung, the amebas travel to the brain by hematogenous spread. *Acanthamoeba* spp. appears to be an opportunistic ameba, but *Naegleria* is not. There are numerous questions that might be answered by fundamental research.

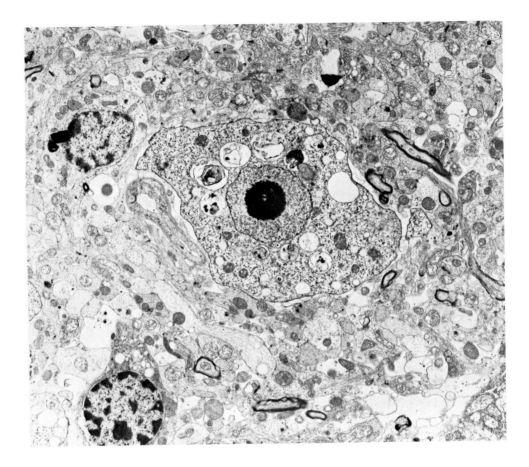

FIGURE 7.   Micrograph of a *Naegleria fowleri* trophozoite within CNS of an infected mouse. There are recognizable myelin sheaths, dendrites, axons, and probably two CNS cells. (Magnification × 3000.)

Immunological, ecological, therapeutic, and prophylactic aspects of free-living and amphizoic amebas are important areas of study for the future using the animal model. Experiments on animal models may be able to answer important questions.

Does pollution of water play a role in the virulence of free-living amebas, and if so, how? Are there other portals of entry into the CNS besides the olfactory and respiratory routes? Is *Vahlkampfia* spp. involved in human disease? What are the precise species of *Acanthamoeba* involved in human and animal disease? Can nonpathogenic *Naegleria* and *Acanthamoeba* spp. become pathogenic? Are the pathogenic strains of free-living amebas mutants and, if so, by what mechanism? Can the higher temperatures of the water (up to 45°C) influence the virulence of free-living amebas? Can free-living and amphizoic amebas act as vectors for viral and bacterial organisms? Can free-living amebas and their proteins produce hypersensitivity or allergic conditions? Are there reservoirs of free-living amebas? Do the free-living amebas produce subclinical or asymptomatic infections? Are free-living amebas dust-borne and transported through the air as cysts? Or, can the trophozoites be infective? What drugs may be effective in preventing *Naegleria* and *Acanthamoeba* spp. infection? Can the disease be cured as soon as it is diagnosed? Is the drinking water contaminated with free-living amebas, particularly in cities or communities where there is no filtration? What is the danger in swimming pools, lakes, and rivers, particularly where the chlorination is insufficient? Is there a change in the virulence of the free-living amebas? What is the

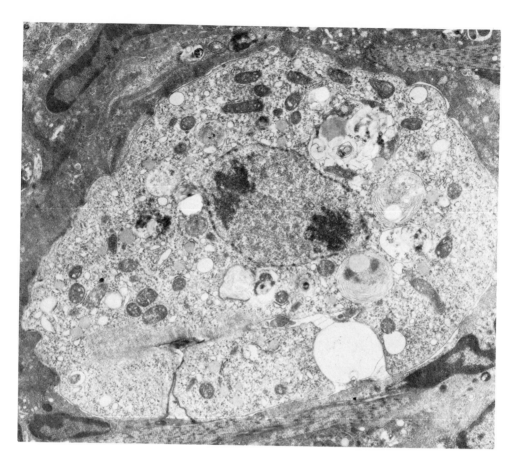

FIGURE 8.   Micrograph of a *Naegleria fowleri* in the olfactory neuroepithelium of an infected mouse. The nucleus is in mitosis. (Magnification × 4520.) (From Martinez, A. J. et al., *Lab. Invest.*, 29(2), 121, 1973. With permission.)

real incidence of "carriers" of nonpathogenic or pathogenic free-living amebas and clinically "asymptomatic" cases? Has the high temperature of the summer months and pollution of the environment lead to genetic changes influencing the virulence of the protozoa? May the hybridoma technology and the production of monoclonal antibodies be useful in the diagnosis of free-living amebic infection? And, finally, how much of our resources should be spent to answer the above questions? In what direction can we go in view of the available evidence of danger to public health? Where does this leave us?

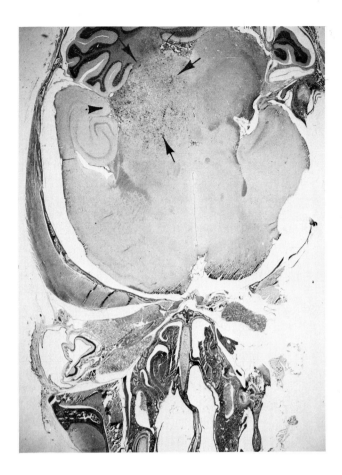

FIGURE 9.   Horizontal section of the head of a mouse 18 days after intranasal inoculation with *Acanthamoeba polyphaga*. There is an inflammatory lesion on the midbrain (arrows). (H & E; magnification × 20.) (From Martinez, A. J., Markowitz, S. M., and Duma, R. J., *J. Infect. Dis.*, 131(6), 692, 1975. Published by The University of Chicago. With permission.)

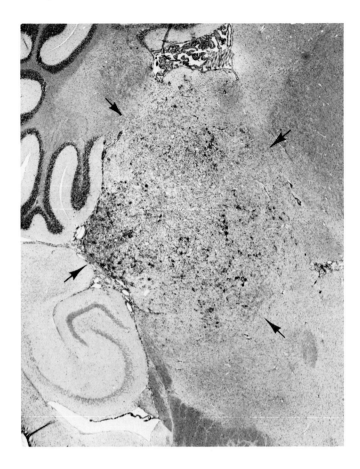

FIGURE 10.   Higher power of the previous photomicrograph demonstrating the diffuse borders and the necrotizing character. (H & E; magnification × 50.)

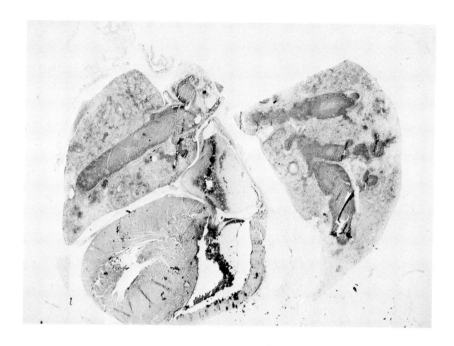

FIGURE 11.   Lungs and heart of a mouse inoculated intranasally 13 days previously with
Acanthamoeba polyphaga. There are foci of bronchopneumonia and extensive consolidation
of the pulmonary parenchyma. (H & E; magnification × 20.)

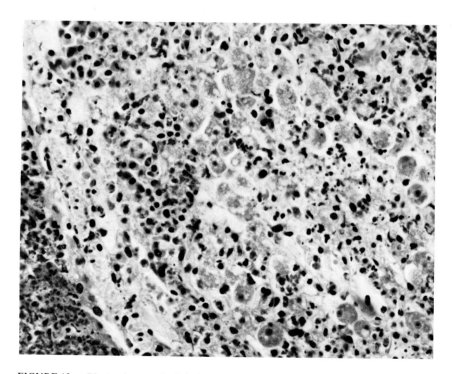

FIGURE 12.   Photomicrograph of the lung of a mouse with *Acanthamoeba polyphaga* pneumonitis. This animal was intranasally inoculated 13 days previously. Amebic trophozoites can
be seen with inflammatory cells and necrotic tissue. (H & E; magnification × 600.) (From
Martinez, A. J., Markowitz, S. M., and Duma, R. J., *J. Infect. Dis.*, 131(6), 692, 1975.
Published by The University of Chicago Press. With permission.)

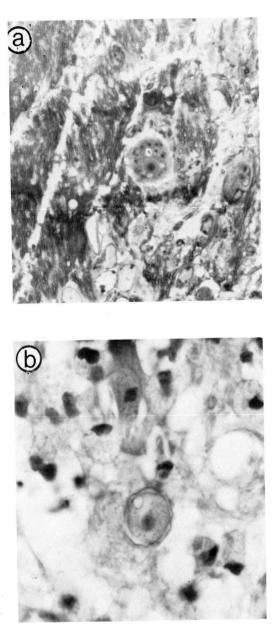

FIGURE 13.   (a) Trophozoite of *Acanthamoeba castellanii* within human CNS. One-micron-thick plastic embedded tissue. (Toluidine blue; magnification × 100.) (b) Cyst of *Acanthamoeba polyphaga* in mouse brain. The wrinkled wall of the cyst is evident. (H & E; magnification × 100.)

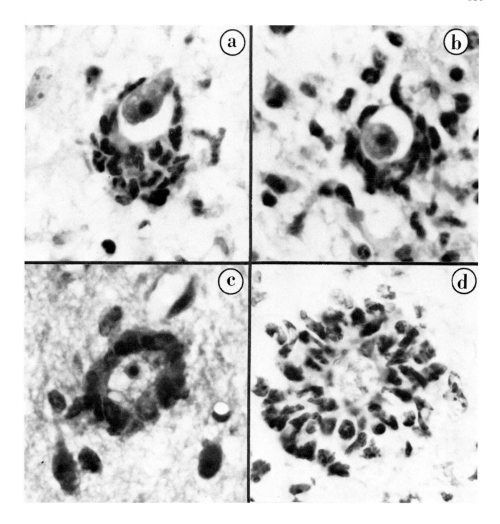

FIGURE 14.   Different stages of phagocytosis of trophozoites of *Acanthamoeba* spp. within infected mouse brain. (a) The amebic trophozoite is surrounded by polymorphonuclear leukocytes and microglia; (b) another amebic trophozoite is encompassed mainly by microglia and lymphocytes; (c) the amebic trophozoite is undergoing vacuolization of the cytoplasm and degenerative changes; (d) this amebic trophozoite is almost destroyed by numerous inflammatory cells, macrophages, and microglia. (H & E; magnification × 500.)

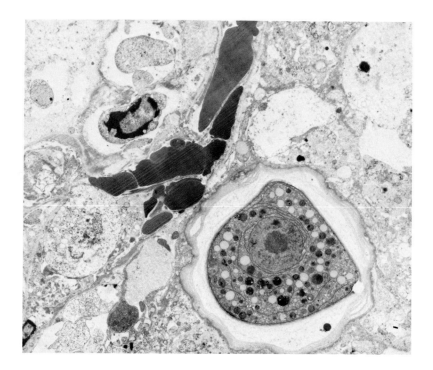

FIGURE 15.   Micrograph of a cyst of *Acanthamoeba castellanii* within the lung of a
mouse. The cyst wall is thick and irregular. (Magnification × 3000.) (From Martinez, A.
J., Markowitz, S. M., and Duma, R. J., *J. Infect. Dis.*, 131(6), 692, 1975. Published by
The University of Chicago Press. With permission.)

141

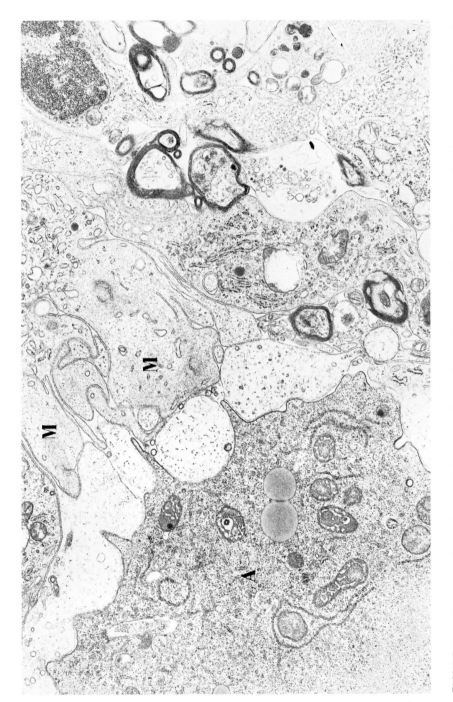

FIGURE 16. Micrograph of a trophozoite of *Acanthamoeba castellanii* in brain mouse. The amebic trophozoite (A) is encompassed by pseudopods of a macrophage (M). Edematous dendrites are located between the surface of the ameba, the "spiked" acanthopodia, and the macrophage. (Magnification × 800.)

REFERENCES

1. Adams, A. C., John, D. T., and Bradley, S. G., Modification of resistance of mice to *Naegleria fowleri* infection, *Infect. Immun.*, 13, 1387, 1976.
2. Bovee, E., Bovee, G., Wilson, D., and Telford, S., Amebiasis of tissues induced in mice and rats by inoculation with *Acanthamoeba* and *Entamoeba* spp., *Am. Zool.*, 1, 439, 1961.
3. Brown, T., Observations by light microscopy on the cytopathogenicity of *Naegleria fowleri* in mouse embryo-cell cultures, *J. Med. Microbiol.*, 11, 249, 1978.
4. Brown, T., Observations by immunofluorescence microscopy and electron microscopy on the cyto-pathogenicity of *Naegleria fowleri* in mouse embryo-cell cultures, *J. Med. Microbiol.*, 12, 363, 1979.
5. Brown, T., Inhibition by amoeba-specific antiserum and by cytochalasin B of the cytopathogenicity of *Naegleria fowleri* in mouse embryo-cell cultures, *J. Med. Microbiol.*, 12, 355, 1979.
6. Carter, R. F., Primary amoebic meningoencephalitis: an appraisal of present knowledge, *Trans. R. Soc. Med. Hyg.*, 66, 193, 1972.
7. Červa, L., Use of fluorescent antibody technique to identify pathogenic *Hartmannellae* in tissues of experimental animals, *Folia Parasitol. (Praha)*, 13, 326, 1966.
8. Červa, L., Intracerebral inoculation of experimental animals in pathogenetical studies of *Hartmannella castellanii*, *Folia Parasitol. (Praha)*, 14, 171, 1967.
9. Červa, L., Intranasal, intrapulmonary and intracardial inoculation of experimental animals with *Hartmannella castellanii*, *Folia Parasitol. (Praha)*, 14, 207, 1967.
10. Červa, L., Growth of the pathogenic A1 strain of *Acanthamoeba castellanii* in the chick embryo, *Folia Parasitol. (Praha)*, 17, 315, 1970.
11. Červa, L., An attempt at selective cultivation of pathogenic *Naegleria gruberi* strains, *J. Protozool.*, Suppl. 18, 44, 1971.
12. Červa, L., Experimental infection of laboratory animals by the pathogenic *Naegleria gruberi*, strain Vitek, *Folia Parasitol. (Praha)*, 18, 171, 1971.
13. Ciplea, A. G., Proca, M., and Milcu, M., Recherches histochiniques sur les modifications des acides nucliques chez la souris blanche infectee par *A. Castellanii*, *Arch. Roum. Pathogie Exp.*, 32, 373, 1973.
14. Culbertson, C. G., Smith, J., and Minner, J. R., *Acanthamoeba:* observations on animal pathogen-icity, *Science*, 127, 1506, 1958.
15. Culbertson, C. G., Smith, J. W., Cohen, H. K., and Minner, J., Experimental infection of mice and monkeys by *Acanthamoeba*, *Am. J. Pathol.*, 35, 185, 1959.
16. Culbertson, C. G., Ensminger, P. W., and Overton, W. M., The isolation of additional strains of pathogenic *Hartmannella* sp. *(Acanthamoeba).* Proposed culture method for application to biological material, *Am. J. Clin. Pathol.*, 43, 383, 1965.
17. Culbertson, C. G., *Hartmannella castellanni* (*Acanthamoeba*). Pathologic lesions produced in exper-imental animals by strains of vary ing degrees of virulence (Abstr.), *Am. J. Clin. Pathol.*, 44(5), 580, 1965.
18. Culbertson, C. G., Holmes, D., and Overton, W., *Hartmannella castellanii* (*Acanthamoeba* sp.). Preliminary report on experimental chemotherapy, *Am. J. Clin. Pathol.*, 43, 361, 1965.
19. Culbertson, C. G., Ensminger, P. W., and Overton, W. N., Experimental Hartmannellosis, *Excerpta Med. Int. Congr. Ser.*, 91, 126, 1965.
20. Culbertson, C. G., Ensminger, P. W., and Overton, W. M., *Hartmannella (Acanthamoeba):* exper-imental chronic granulomatous brain infections produced by new isolates of low virulence, *Am. J. Clin. Pathol.*, 46, 305, 1966.
21. Culbertson, C. G., Ensminger, P. W., and Overton, W. M., Pathogenic *Naegleria* sp. Study of a strain isolated from human cerebrospinal fluid, *J. Protozool.*, 15, 353, 1968.
22. Culbertson, C. G., The pathogenicity of soil amebas, *Ann. Rev. Microbiol.*, 25, 231, 1971.
23. Culbertson, C. G., Ensminger, P. W., and Overton, W. M., Amebic cellulocutaneous invasion by *Naegleria aerobia* with generalized visceral lesions after subcutaneous inoculation, *Am. J. Clin. Pathol.*, 57, 375, 1972.
24. Cursons, R. T. M. and Brown, T. J., Use of cell cultures as an indicator of pathogenicity of free-living amoebae, *J. Clin. Pathol.*, 31, 1, 1978.
25. De Jonckheere, J., Differences in virulence of *Naegleria fowleri*, *Pathol. Biol.*, 27, 453, 1979.
26. DeJonckheere, J. F., Aerts, M., and Martinez, A. J., Naegleria australiensis: experimental menin-goencephalitis in mice, *Trans. R. Soc. Trop. Med. Hyg.*, 77, 712, 1983.
27. Dempe, S., Martinez, A. J., and Janitschke, K., Subacute and chronic meningoencephalitis in mice after experimental infection with a strain of *Naegleria fowleri* originally isolated from a patient, *Infection*, 10, 5, 1982.
28. Diffley, P., Skeels, M. R., and Sogandares-Bernal, F., Delayed type hypersensitivity in guinea pigs infected subcutaneously with *Naegleria fowleri*, Carter, *Zentralbl. Parasitenkunde*, 49, 133, 1976.

29. Ensminger, P. W. and Culbertson, C. G., *Hartmannella (Acanthamoeba)*. Experiments in preservation and detection of trophozoites and amebic cysts in infected tissue, *Am. J. Clin. Pathol.*, 46, 496, 1966.

30. Ensminger, P. and Culbertson, C., *Hartmannella (Acanthamoeba)* experiments in preservation and detection of trophozoites and amebic cysts in infected tissue, *Tech. Bull. Regist. Med. Technol.*, 36, 234, 1966.

31. Ferrante, A. and Thong, Y. H., Unique phagocytic process in neutrophil-mediated killing of *Naegleria fowleri, Immunol. Lett.*, 2, 37, 1980.

32. Ferrante, A., Experimental pneumonitis induced by *Naegleria fowleri* in mice, *Trans. R. Soc. Trop. Med. Hyg.*, 75(6), 907, 1981.

33. Ferrante, A., Comparative sensitivity of *Naegleria fowleri* to amphotericin B and amphotericin B methyl ester, *Trans. R. Soc. Trop. Med. Hyg.*, 76, 476, 1982.

34. Ferrante, A., Rowan-Kelly, B., and Thong, Y. H., In vitro sensitivity of virulent *Acanthamoeba culbertsoni* to a variety of drugs and antibiotics, *Int. J. Parasitol.*, 14, 53, 1984.

35. Haggerty, R. M. and John, D. T., Innate resistance of mice to experimental infection with *Naegleria fowleri, Infect. Immun.*, 20, 73, 1978.

36. Haggerty, R. M. and John, D. T., Factors affecting the virulence of *Naegleria fowleri* for mice, *Proc. Helminthol. Soc. Wash.*, 47(1), 129, 1980.

37. Haggerty, R. M. and John, D. T., Serum agglutination and immunoglobulin levels of mice infected with *Naegleria fowleri, J. Protozool.*, 29, 117, 1982.

38. Holbrook, T. and Parker, B., *Naegleria fowleri* in chick embryos. Effects of embryo age and incubation temperature, and the infectivity of embryo-derived amebae for mice, *Am. J. Trop. Med. Hyg.*, 28(6), 984, 1979.

39. Holbrook, T. W., Boackle, R. J., Parker, B. W., and Vesely, J., Activation of the alternative complement pathway by *Naegleria fowleri, Infect. Immun.*, 30, 58, 1980.

40. John, D. T., Weik, R., and Adams, A., Immunization of mice against *Naegleria fowleri* infection, *Infect. Immun.*, 16, 817, 1977.

41. John, D. T. and Martinez, A. J., Experimental primary amebic meningoencephalitis *(Naegleria)* in mice following intravenous inoculation, *J. Protozool.*, 22, 39A, 1977.

42. Karr, S. L. and Wong, M. W., Susceptibility of inbred mice to intranasal infection with high and low virulence isolates of *Naegleria fowleri* HB-1 strain, *Trans. R. Soc. Trop. Med. Hyg.*, 74, 127, 1980.

43. Lal-Altaf, A. and Garg, N. K., *Hartmannella culbertsoni*: biochemical changes in the brain of the meningoencephalitic mouse, *Exp. Parasitol.*, 48, 331, 1979.

44. Maitra, S. C., Krishna Prasad, B. N., Das, S. R., and Agarwala, S. C., A study of *Naegleria aerobia* by electron microscopy, *Trans. R. Soc. Trop. Med. Hyg.*, 68, 56, 1974.

45. Maitra, S. C., Krishna Prasad, B. N., Das, S. R., and Agarwala, S. C., Ultrastructural differences of *Hartmannella culbertsoni*, Singh and Das, 1970, in mouse brain and under different cultural conditions, *Trans. R. Soc. Trop. Med. Hyg.*, 68, 229, 1974.

46. Markowitz, S. M., Sobieski, T., Martinez, A. J., and Duma, R., Experimental *Acanthamoeba* infections in mice pretreated with methylprednisolone or tetracycline, *Am. J. Pathol.*, 92, 733, 1978.

47. Martinez, A. J., Nelson, E. C., Duma, R., Jones, M., Huff, S., and Rosenblum, W. I., Experimental amebic meningoencephalitis in mice, *J. Neuropathol. Exp. Neurol.*, 31, 173, 1972.

48. Martinez, A. J., Duma, R. J., Nelson, E. C., and Moretta, F. L., Experimental *Naegleria* meningoencephalitis in mice. Penetration of the olfactory mucosal epithelium by *Naegleria* and pathologic changes produced: a light and electron microscope study, *Lab. Invest.*, 29, 121, 1973.

49. Martinez, A. J., Markowitz, S. M., and Duma, R. J., Experimental pneumonitis and encephalitis caused by *Acanthamoeba* in mice: pathogenesis and ultrastructural features, *J. Infect. Dis.*, 131, 692, 1975.

50. Martinez, A. J., Nelson, E. C., and Duma, R. J., Animal model: primary amebic *(Naegleria)* meningoencephalitis in mice, *Am. J. Pathol.*, 73, 545, 1973.

51. Martinez, A. J., Nelson, E. C., Jones, M. M., Duma, R. J., and Rosenblum, W. I., Experimental *Naegleria* meningoencephalitis in mice: an electron microscopy study, *Lab. Invest.*, 25, 465, 1971.

52. Martinez, A. J., Primary amebic meningoencephalitis. Model No. 48, supplemental update, 1982, in Handbook: Animal Models of Human Disease, Fasc. 11, Capen, C. C., Hacke, D. B., Jones, T. C., and Migaki, G., Eds., Registry of Comparative Pathology, Armed Forces Institute of Pathology, Washington, D.C., 1982.

53. May, R. G. and John, D. T., Intravenous infection of mice with *Naegleria fowleri, Folia Parasitol. (Praha)*, 29, 201, 1982.

54. May, R. G. and John, D. T., Transmission of *Naegleria fowleri* between mice, *J. Parasitol.*, 69, 249, 1983.

55. Misra, R. and Sharma, A. K., Pathogenicity in mice of strains of *Acanthamoeba culbertsoni* (Singh & Das, 1970) and *A. rhysodes* (Singh, 1952) from Indian soils and the present status of the genera *Hartmannella* and *Acanthamoeba, Ind. J. Parasitol.*, 4, 135, 1980.

56. Pernin, P. and Riany, A., Étude comparative du pouvoir pathogène expérimental de souches d' *Acanthamoeba, Ann. Parasitol. (Paris),* 55(5), 491, 1980.

57. Pernin, P., Riany, A., and Grimaud, J. A., Étude ultrastructurale de la méningoencéphalite expérimentale provoqué epar une souche d' *Acanthamoeba* isolée de piscine, *Protistologica,* 15(3), 307, 1979.

58. Phillips, B. P., Naegleria: another pathogenic ameba. Studies in germ free guinea pigs, *Am. J. Trop. Med. Hyg.,* 23, 850, 1974.

59. Proca, M. and Ciplea, A. G., Infection experimentale a *A. castellani* chez les souris blanches, *Arch. Roum. Pathol. Exp. Microbiol.,* 32, 457, 1973.

60. Proca-Ciobanu, M. and Ciplea, A., Contributions to the study of the experimental infection with *Acanthamoeba castellanii* with white mice: histopathological and histochemical data, in Progress in Protozoology (Clermont-Ferrand) 330, 4th Int. Congr. on Protozoology, September 2 to 10, 1973.

61. Proca-Ciobanu, M. and Ionescu, M. D., Ultrastructural changes of *Acanthamoeba culbertsoni* in the brain of the experimentally infected white mouse, *Arch. Roum. Pathol. Exp. Microbiol.,* 34, 329, 1975.

62. Rowan-Kelly, B., Ferrante, A., and Thong, Y. H., Activation of complement by *Naegleria, Trans. R. Soc. Trop. Med. Hyg.,* 74, 333, 1980.

63. Rowan-Kelly, B., Ferrante, A., and Thong, Y. H., The chemotherapeutic value of sulphadiazine in treatment of *Acanthamoeba* meningoencephalitis in mice, *Trans. R. Soc. Trop. Med. Hyg.,* 76, 636, 1982.

64. Schlaegel, T. G., Jr. and Culbertson, C. G., Experimental Hartmannella optic neuritis and uveitis, *Ann. Ophthalmol.,* p. 103, 1972.

65. Simpson, C. F., Willaert, E., Neal, F. C., Stevens, A. R., and Young, M. D., Experimental *Naegleria fowleri* meningoencephalitis in sheep: light and electron microscopic studies, *Am. J. Vet. Res.,* 43(1), 154, 1982.

66. Singh, B. N. and Das, S. R., Intranasal infection of mice with flagellate stage of *Naegleria aerobia* and its bearing on the epidemiology of human meningoencephalitis, *Curr. Sci.,* 41, 625, 1972.

67. Singh, B. N. and Hanumaiah, V., Temperature tolerance of free-living amoebae and their pathogenicity to mice, *Ind. J. Parasitol.,* 1, 71, 1977.

68. Thong, Y. H., Shepherd, C., Ferrante, A., and Rowan-Kelly, B., Protective immunity to *Naegleria fowleri* in experimental amebic meningoencephalitis, *Am. J. Trop. Med. Hyg.,* 27, 238, 1978.

69. Thong, Y. H., Ferrante, A., Shepherd, C., and Rowan-Kelly, B., Resistance of mice to *Naegleria* meningoencephalitis transferred by immune serum, *Trans. R. Soc. Trop. Med. Hyg.,* 72, 650, 1978.

70. Thong, Y. H., Rowan-Kelly, G., and Ferrante, A., Pyrimethamine in experimental amoebic meningoencephalitis, *Aust. Paediatr. J.,* 14, 177, 1978.

71. Thong, Y. H., Ferrante, A., Rowan-Kelly, B., and O'Keefe, D., Immunization with live amoebae, amoebic lysate and culture supernatant in experimental *Naegleria* meninoencephalitis, *Trans. R. Soc. Trop. Med. Hyg.,* 74, 570, 1980.

72. Thong, Y. H., Ferrante, A., Rowan-Kelly, B., and O'Keefe, D. E., Immunization with culture supernatant in experimental amoebic meningoencephalitis, *Trans. R. Soc. Trop. Med. Hyg.,* 73, 684, 1979.

73. Thong, Y. H., Carter, R. F., Ferrante, A., and Rowan-Kelly, B., Site of expression of immunity to *Naegleria fowleri* in immunized mice, *Parasite Immunol.,* 5, 67, 1983.

74. Visvesvara, G. S. and Callaway, C. S., Light and electron microscopic observations on the pathogenesis of *Naegleria fowleri* in mouse brain and tissue culture, *J. Protozool.,* 21, 239, 1974.

75. Willaert, E. and Stevens, A. R., Experimental pneumonitis induced by *Naegleria fowleri* in mice, *Trans. R. Soc. Trop. Med. Hyg.,* 74, 779, 1980.

76. Wilson, D., Bovee, E., Bovee, G., and Telford, S., Induction of amebiasis of tissues of white mice and rats by subcutaneous inoculation of small free-living, inquiline, and parasitic with associated coliform bacteria, *Exp. Parasitol.,* 21, 277, 1967.

77. Wong, M. W., Karr, S. L., and Balamuth, W. B., Experimental infections with pathogenic free-living amebae in laboratory primate hosts. I. (A) A study in susceptibility to *Naegleria fowleri, J. Parasitol.,* 61, 199, 1975.

78. Young, M. D., Willaert, E., Neal, F. C., Simpson, C. F., and Stevens, A. R., Experimental infection of sheep with *Naegleria fowleri* of human origin, *Am. J. Trop. Med. Hyg.,* 29, 476, 1980.

Chapter 10

FREE-LIVING AND AMPHIZOIC AMEBAS IN ANIMAL HOSTS

I. INTRODUCTION

The ubiquitous presence of free-living amebas in nature suggests the possibility that they may live in domestic and wild animals as parasites or that they may infect and produce disease in them. Therefore, the public health implications, as well as interest in these protozoa by medical and veterinary science, are obvious.[27,52,53]

II. MAMMALS

Naturally occurring infections due to *Acanthamoeba* spp. have been reported in several mammals: buffalo, cattle, bull, sheep, gibbon, dog, and wallaroo or kangaroo[1,10,11,20,28,38,40] (Table 1). Some of these animals died of unsuspected amebic infection. In others, ulcerations and erosions of colonic mucosa and granulomas of nasal mucosa were found during autopsy. Apparently, these lesions were contributory to the cause of death. Free-living amebic trophozoites and cysts were found at post-mortem involving different internal organs: lungs, heart, liver, kidney, or brain.

Domestic animals (dogs, pigs, horses, cats, and rabbits) have been found carrying free-living amebas.[5,8,19,21,25,33]

Species of the genus *Entamoeba* (i.e., *E. bovis, E. histolytica)* are known to be parasites in the gastrointestinal tracts of numerous animals. The mammalian hosts may remain healthy or develop lesions; they include horses, cattle, hogs, sheep, goats, gnus, rabbits, guinea pigs, ground squirrels, monkeys, rats and mice.[44,53] At the San Diego Zoological Park free-living or amphizoic amebas were found to be associated with the death of 20 of the 2561 zoo animals autopsied there. Of the 20 cases of amebiasis, at least 2 were due to *Acanthamoeba* spp. This gives an incidence rate of amebiasis of at least 0.78%. However, probably the direct cause of death in some of these animals was not directly related to the amebas since other pathological processes were found.[8,20,21]

Wild animals (muskrat, oppossum, raccoon, and cotton rat) may also be susceptible to free-living amebic infection. These animals, when intranasally inoculated with pathogenic *Naegleria fowleri* or *Acanthamoeba* spp., may develop amebic disease. Some wild animals, like the muskrat, that live in the same aquatic environment as the free-living amebas also appear to be susceptible to the amebic infection.[24] These animals, despite the frequent exposure to free-living amebas, do not appear to have or develop immunity to the infection and probably are not entirely resistant to the infection.

Spontaneous infection due to *Acanthamoeba* spp. was found in a beaver *(Castor canadensis)*. This beaver was captured in Canada and kept as a pet in Switzerland. Amebic trophozoites were found in the kidneys, duodenum, and jejunum, liver, heart, and skeletal muscles. The amebic trophozoites were found within foci of necrosis.[31]

A white-cheeked gibbon autopsied at the San Diego Zoo revealed a brain abscess due to free-living amebas, and a Nelson's bighorn sheep, also from the same zoo, had a granulomatous lesion in the nasal mucosa due to free-living amebas.[20]

III. BIRDS AND FOWL

Pigeons, turkeys, and other fowl may harbor and act as natural hosts of amphizoic and free-living amebas. In fact, free-living amebas have been found in their gastrointestinal tracts.[50] Spontaneous free-living amebic infections in birds and fowl have not been reported, however.

Table 1

SPONTANEOUS INFECTIONS WITH FREE-LIVING AND
AMPHIZOIC AMEBAS IN ANIMALS

Country/state	Host	Tissue infected	Ref.
India	Water buffalo	Lungs	10
Vietnam	German shepherd dog	Heart, lungs, liver, pancreas (brain not examined)	1
Azores	Bull (Holstein)	Lung	28
California	Nelson bighorn sheep	Nasal mucosa; olfactory mucosa; granulomatous lesions	8, 20
	White-cheeked gibbon	Brain abscess	21
Florida	Greyhound dog	Lungs	19
Switzerland	Beaver: *Castor canadensis*	Heart, liver, kidneys, GI tract, and skeletal muscles	31
U.S.	Goldfish: *Carassius auratus*	Kidney, liver, brain, meninges, swim bladder, skeletal muscles	51
Italy	Several species of fish	Intestinal mucosa	43
Germany	*Iguana iguana*	GI tract, CNS, eyelids	12

Modified from Visvesvara, G. S., The public health importance and disease potential of small free-living amebae,[28-45] presented at the 2nd Int. Conf. on the Biology and Pathogenicity of Small Free-Living Amoebae, University of Florida, Gainesville, March 23 to 25, 1980, 28.

IV. REPTILES AND AMPHIBIANS

Free-living amebas (other than *Acanthamoeba* spp. and *Naegleria fowleri*) may infect cold-blood animals[2-4,6,7,22,36,37] (Table 2).

Endolimax ranarum and *Entamoeba invadens* have been found in cold-blooded animals such as snakes, iguanas, frogs, and turtles.[13-17] *E. invadens* has also been found in an *Iguana iguana* that died of disseminated amebiasi. (Figure 1). Most of the internal organs, including the liver, heart, and brain, were affected. The amebic disease also severely involved the skin and eyelids. Both trophozoites and cysts were present in the lesions. However, the gastrointestinal tract was not involved. A large brain abscess was noted to contain numerous amebic trophozoites and a few cysts (Figure 2). Multiple small amebic foci were also present within the myocardium.[12]

V. FISH AND CRUSTACEANS

Isolation of strains of *Acanthamoeba* spp., *Naegleria*, and *Vahlkampfia* have been reported in fresh-water fish, and, in some instances, killing large numbers of fish. *Acanthamoeba* spp. and *Vahlkampfia* spp. were isolated from either the gills, intestines, or the peritoneal cavity of *Tilapia aurea* and different types of bass. Trophozoites and cysts were present within pathologic lesions.[43,48,49,51]

Sawyer and associates described seasonal epizootic amebiasis in rainbow trout in Italy and succeeded in growing the etiological ameba in bacteria-agar media. They found pathological lesions in the kidney, spleen, liver, and peritoneum of this trout.[42]

Crabs and lobsters have also been reported to harbor or be infected with free-living amebas.[41,45,46] *Paramoeba perniciosa* produced phagocytic nodules in rock blue and gray crabs and in lobsters.[41-43,45] Both trophozoites and cystic forms were found. *Vahlkampfia patuxent* n. sp. have been reported parasitizing oysters[23] and other mollusks.[18]

Table 2
SPONTANEOUS INFECTIONS WITH OTHER FREE-LIVING AND AMPHIZOIC AMEBAS

Host	Agent: isolated/suspected	Ref.
Snails (*Bulinus globosus* and *Biomphalaria pallida*)	*Hartmannella biparia, H. quadriparia*	32,39
Oyster (*Carassostrea commercialis*)	*Hartmannella tahitiensis*	6,23
Blue and gray crabs *(Callinectes sapidus)*	*Paramoeba perniciosa*	45,46
Rock crab (*Cancer irroratus*) Lobster (*Homarus americansus*)	*Paramoeba perniciosa*	39,41
Rainbow trout	*Vexillifera bacillipedes*	43
Blue tilapia and bass	*Acanthamoeba* spp.	48
Iguana iguana	*Entamoeba invadens*	12—14
Sporocysts of *Schistosoma mansoni*	*Nucleria* sp.	32,35

Modified from Visvesvara, G. S., The public health importance and disease potential of small free-living amebae, presented at the 2nd Int. Conf. on the Biology and Pathogenicity of Small Free-Living Amoebae, University of Florida, Gainesville, March 23 to 25, 1980, 28.

FIGURE 1. *Iguana iguana* showing the extensive ulceration of the eyelids. (Courtesy of Profesor W. Franz and Dr. U. Bachman.)

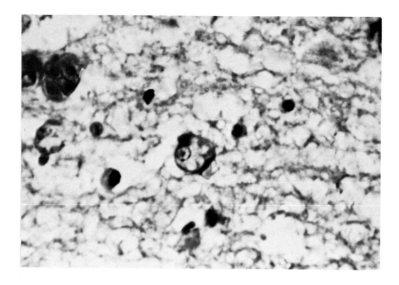

FIGURE 2.   Trophozoite of *Entamoeba invadens* within a cerebral abscess. (H &
E; Magnification × 800.) (From Frank, W., Z. *Tropen Med. Parasitol.*, 17(3), 285,
1966. With permission.)

VI. MOLLUSKS

In addition to *Acanthamoeba* and *Naegleria fowleri,* other free-living amebas belonging to such genera as *Hartmannella, Paramoeba, Vexillifera,* and *Nuclearia* have
been implicated in causing disease in a variety of invertebrates. The genus *Acanthamoeba* has been found living in the sea waters of the North Atlantic, Florida straits, and
other offshore marine environment.[9,42]

Epidemics have been reported in oysters due to *H. tahitiensis* sp. no.[6] Fresh-water
mollusks and other snails *(Bulinus globusus* and *Biomphalaria pallida)* were found to
be infected by *Hartmannelid* amebas.[39] These amebas impaired the growth and reproduction of the host snails.[32,35,39]

VII. INSECTS

Grasshoppers may also be carriers or infected by *Acanthamoeba.*[26] Trophozoites
and cysts were found within the epithelial cells of the midgut.[47] *Vahlkampfia* amebas
have been found living in the intestines of termites[29] and *Endamoeba blattae* may live
as a parasite in the colon of cockroaches.[27,30]

REFERENCES

1. Ayers, K. J., Billups, L. H., and Garner, F. M., *Acanthamoebiasis* in a dog, *Vet. Pathol.*, 9, 221,
   1972.
2. Bosch, J. and Deichsel, G., Morphologische untersuchungen an pathogenen und potentiell pathogenen amoeben der typen *"Entamoeba"* und *"Hartmannella Acanthamoebe"* aus Reptilien, *Z. Parasitenkunde,* 40, 107, 1972.
3. Bovee, E. C., Wilson, D. E., and Telford, S. R., Some amebas and amoeboflagellates inquilinic in
   Florida reptiles, *J. Protozool.*, 8 (Suppl.), 15, 1961.
4. Bovee, E. C., Wilson, D. E., and Telford, S. R., Entozoic amebas from feral reptiles, *Am. Zool.*, 1,
   439, 1961.

5. Černa, L., Spontaneous occurrence of antibodies against pathogenic amoebae of the limax group in domestic animals, *Folia Parasitol. (Praha),* 28, 105, 1981.
6. Cheng, T. C., *Hartmannella tahitiensis* sp. n., an amoeba associated with mass mortalities of the oyster *Crassostrea commercialis* in Tahiti, French Polynesia, *J. Invertebrate Pathol.,* 15, 405, 1970.
7. Ciurea-Van Saanen, M. D., Isolation of free-living amoebae from cold blooded animals, *Schweiz. Arch. Tierheilkd.,* 122, 6543, 1980.
8. Culbertson, C. G., personal communication for Griner, La., *Annu. Rev. Microbiol.,* 25, 231, 1971.
9. Davis, P. G., Caron, D. A., and McN. Sieburth, J., Oceanic amoebae from the North Atlantic: Culture, distribution and taxonomy, *Trans. Am. Microsc. Soc.,* 97, 73, 1978.
10. Dwivedi, J. N. and Singh, C. M., Pulmonary lesions in an Indian buffalo associated with *Acanthamoeba* sp., *Ind. J. Microbiol.,* 3, 31, 1965.
11. Eyles, D. E., Jones, F. E., Jumper, J. R., and Drinnon, V. P., Amebic infections in dogs, *J. Parasitol.,* 40, 163, 1954.
12. Frank, W., Generalisierte amöbiasis ohne Darmsymptome bei einem Leguan *(Iguana iguana)* (Reptilia, *Iguanidae),* hervorgerufen durch *Entamoeba invadens* (Protozoa amoebozoa), *Z. Tropenmed. Parasitol.,* 17, 285, 1966.
13. Frank, W. and Bosch, I., Isolierung von amoeben des typs "*Hartmannella Acanthamoeba*" und *Naegleria* aus Kaltblutern, *Z. Parasitenkunde,* 40, 139, 1972.
14. Frank, W., *Limax*-amoeba from cold-blooded vertebrates, *Ann. Soc. Belge Med. Trop.,* 54, 343, 1974.
15. Franke, E. and Mackiewicz, J., Isolation of *Acanthamoeba* and *Naegleria* from the intestinal contents of freshwater fishes and their potential pathogenicity, *J. Parasitol.,* 68, 164, 1982.
16. Geiman, Q. M. and Ratcliffe, H. L., Morphology and life-cycle of an amoeba producing amebiases in reptiles, *Parasitology,* 28, 208, 1936.
17. Geiman, Q. M. and Wichtermann, R., Intestinal protozoa from Galapagos tortoises (with descriptions of three new species), *J. Parasitol.,* 23, 331, 1937.
18. Getz, L. L., An attempt to infect mollusks with *Acanthamoeba* sp., *J. Parasitol.,* 47, 842, 1961.
19. Griffin, J. L., Pathogenic free-living amoebae, in *Parasitic Protozoa 2,* Krier, J. P., Ed., Academic Press, New York, 1978.
20. Griner, L. A. and Monroe, L. S., Amebiasis of mammals at San Diego Zoo, paper presented at the Centennial Symp. on Science and Research, Philadelphia Zoo, November 11 to 14, 1974.
21. Griner, L. A., *Pathology of Zoo Animals,* Zoological Society, San Diego Press, 1983, 361.
22. Hawes, R. S. J., A *Limax* amoeba from the rectum of the grass snake, *Natrix natrix* as a facultative aerobe in vitro, *Nature,* 175, 779, 1955.
23. Hogue, M. J., Studies on the life history of *Vahlkampfia patuxent* n.sp. parasitic in the oyster, with experiments regarding its pathogenicity, *Am. J. Hyg.,* 1, 321, 1921.
24. John, D. T., personal communication, 1983.
25. Kadlec, V., The occurrence of amphizoic amebae in domestic animals, *J. Protozool.,* 25, 235, 1978.
26. King, R. L. and Taylor, A. B., *Malpighamoeba locustae* n. sp. (Amoebidae), a Protozoan parasite in the malpighian tubes of grasshoppers, *Trans. Am. Microsc. Soc.,* 55, 6, 1936.
27. Kudo, R. R., Observations on *Endamoeba blattae,* *Am. J. Hyg.,* 6, 139, 1926.
28. McConnell, E. E., Garner, F. M., and Kirk, J. H., Hartmannellosis in a bull, *Pathol. Vet.,* 5, 1, 1968.
29. deMello, I. Froilano, Um amebiano do genero *Wahlkampfia* parasita do intestino de un coptotermes Indiano, *Rev. Soc. Mex. Hist. Nat.,* 10, 53, 1949.
30. Morris, S., Studies of *Endamoeba blattae, J. Morphol.,* 59, 225, 1936.
31. Muller, R., Nachweis von amoeben in herdformigen Organnekrosen bei einem jungen kanadischen Biber *(Castor canadensis), Acta Trop. (Basel),* 30, 373, 1973.
32. Newton, W. L., The comparative tissue reaction of two strains of *Australorbis glabratus* to infection with *Schistosoma mansoni, J. Parasitol.,* 38, 362, 1952.
33. Noble, F. A. and Noble, E. R., *Entamoeba* in farm animals, *J. Parasitol.,* 38, 571, 1952.
34. Noble, G. A., Coprozoic protozoa from Wyoming mammals, *J. Protozool.,* 5, 69, 1958.
35. Owczarzak, A., Stibbs, H. H., and Bayne, C. J., The destruction of *Schistosoma mansoni* mother sporocysts in vitro by amoeba isolated from *Biomphalaria glabrata:* an ultrastructural study, *J. Invertebrate Pathol.,* 35(1), 26, 1980.
36. Page, F. C., *Rosculus ithacus* Hawes, 1963 *(Amoebida, Flabelluidae)* and the amphizoic tendency in amoebae, *Acta Protozool.,* 13, 143, 1974.
37. Ratcliffe, H. L. and Geiman, Q. M., Spontaneous and experimental amebic infection in reptiles, *Arch. Pathol.,* 25, 160, 1938.
38. Rees, C. W., Pathogene's of intestinal amibiasis in kittens, *Arch. Pathol.,* 7, 1, 1929.
39. Richards, C. S., Two new species of *Hartmannella* amebae infecting freshwater mollusks, *J. Protozool.,* 15, 651, 1968.

40. Roberts, E. D., Williams, J. C., and Pirie, G., Naturally occurring gastric amebiasis in the Wallaroo, *Vet. Pathol.*, 10, 323, 1973.

41. Sawyer, T. K., Two new crustacean hosts for the parasitic amoeba, *Paramoeba perniciosa, Trans. Am. Microsc. Soc.*, 95, 271, 1976.

42. Sawyer, T. K., Visvesvara, G. S., and Harke, B. A., Pathogenic amoebas from brackist and ocean sediments with a description of *Acanthamoeba hatchetti* n.sp., *Science*, 196, 1324, 1977.

43. Sawyer, T. K., Ghittino, P., Andruetto, S., Pernin, P., and Pussard, M., *Vexillifera bacillipedes*, page 1969, an amphizoic amoeba of hatchery rainbow trout in Italy, *Trans. Am. Microsc. Soc.*, 97, 596, 1978.

44. Simitzis-Le Flohic, A. M. and Chastel, C., Small wild mammals: vectors of free-living amoebae?, *Med. Trop.*, 42, 275, 1982.

45. Sprague, V. and Beckett, R. L., A disease of blue crabs *(Callinectes sapidus)* in Maryland and Virginia, *J. Invertebrate Pathol.*, 8, 287, 1966.

46. Sprague, V., Beckett, R. L., and Sawyer, T. K., A new species of *Paramoeba (Amoebida, Paramoebidae)* parasitic in the crab *Callinectes sapidus, J. Invertebrate Pathol.*, 14, 167, 1969.

47. Taylor, A. B. and King, R. L., Further studies on the parasitic amoebae found in grasshoppers, *Trans. Am. Microsc. Soc.*, 56, 172, 1937.

48. Taylor, P. W., Isolation and experimental infection of free-living amebae in freshwater fishes, *J. Parasitol.*, 63, 232, 1977.

49. Trust, T. J. and Bartlett, K. H., Occurrence of potential pathogens in water containing ornamental fishes, *Appl. Microbiol.*, 28, 35, 1974.

50. Tyzzer, E. E., Amoebae of the caeca of the common fowl and the turkey, *J. Med. Res.*, 41, 199, 1920.

51. Voelker, F. A., Anver, M. R., McKee, A. E., Casey, H. W., and Brenniman, G. R., Amebiasis in goldfish, *Vet. Pathol.*, 14, 247, 1977.

52. Wilhelm, W. E. and Anderson, J. H., *Vahlkampfia labospinosa* (Craig, 1912) Craig 1913: rediscovery of a coprozoic ameba, *J. Parasitol.*, 57, 1378, 1971.

53. Wilson, D. E., Bovee, E. C., Bovee, G. J., and Telford, S. R., Jr., Induction of amebiasis in tissues of white mice and rats by subcutaneous inoculation of small free-living inquilinic and parasitic amebas with associated coliform bacteria, *Exp. Parasitol.*, 21, 277, 1967.

INDEX